Nonlocal Mechanical Behaviour of Functionally Graded Composite Nanostructures

The intricate domain of composite structures, particularly functionally graded nanostructures, presents a fascinating area of study with immense potential for scientific and industrial advancements. This book delves into the unique properties and applications of these advanced materials, providing an in-depth understanding of the continuum theories that underpin functionally graded nanostructures. The authors explore both theoretical foundations and practical insights into the modelling and numerical simulations of the mechanical behaviour of functionally graded materials and structures.

- Provides a thorough exploration of the wide array of modelling techniques and computational methods pivotal for studying functionally graded nanostructures
- Offers unique insights into the behaviour of functionally graded materials, allowing for a multifaceted exploration of their properties and applications
- Covers the influence of material gradation on mechanical performance and the effects of environmental factors like temperature and humidity
- Combines theoretical foundations with practical insights
- Discusses challenges associated with manufacturing and testing these advanced materials

By presenting a detailed and nuanced view of functionally graded nanostructures, this book equips readers in materials engineering and related fields with the knowledge and tools needed to understand the complex interactions and behaviours of these materials under various conditions.

Chinika Dangi is an Assistant Professor in the Department of Mathematics at Amrita Vishwa Vidyapeetham, Chennai. She was a Postdoctoral Fellow in the Department of Aerospace Engineering at the Indian Institute of Science, Bangalore. She earned a PhD at the Indian Institute of Technology, Roorkee, and an MSc at the Indian Institute of Technology, Delhi.

Susmita Naskar is a Senior Lecturer leading the Engineered Materials and Structures Lab in the School of Engineering at the University of Southampton. Her research group works in multiscale structural mechanics and multiphysics analysis, focusing on engineered materials and structures involving the intersection of additive manufacturing, material characterization through computational design, and experiments in engineering. Her group is developing methods to address long-term horizon problems and challenges of coupling scales in complex multiphysics material modelling. In addition, she is working on advanced manufacturing techniques that are relevant to the fabrication of engineered materials.

Nonlocal Mechanical Behaviour of Functionally Graded Composite Nanostructures

Chinika Dangi and Susmita Naskar

CRC Press
Taylor & Francis Group
Boca Raton London New York

CRC Press is an imprint of the
Taylor & Francis Group, an **informa** business

Designed cover image: © 2026 Shutterstock

First edition published 2026
by CRC Press
2385 NW Executive Center Drive, Suite 320, Boca Raton FL 33431

and by CRC Press
4 Park Square, Milton Park, Abingdon, Oxon, OX14 4RN

CRC Press is an imprint of Taylor & Francis Group, LLC

ISBN: 978-1-032-95229-1 (hbk)
ISBN: 978-1-032-95230-7 (pbk)
ISBN: 978-1-003-58380-6 (ebk)

DOI: 10.1201/9781003583806

Typeset in Palatino
by codeMantra

Contents

Preface viii

1. Introduction to Nonlocal Continuum Theory and Composite Nanostructures 1
1.1 Overview 1
1.2 Numerical Techniques 5
1.3 Nonlocal Characteristics of BD-FG Nanostructures 6
1.3.1 Organization of the Book 6
References 7

2. Thermo-Mechanical Model of the Vibration Behaviour of BD-FG Nonuniform Nanobeams 17
2.1 Overview 17
2.2 Mathematical Framework 18
2.2.1 Temperature-Dependent Material Properties of BD-FG Nanobeam 19
2.2.2 Eringen's Nonlocal Elasticity Theory 20
2.2.3 Energy Relations 21
2.2.4 Nonlocal Governing Equations of Motion 22
2.3 Method of Solution 24
2.4 Results and Discussion 26
2.4.1 Convergence Study and Validation 26
2.4.2 Vibration Analysis 27
2.5 Summary 32
References 33

3. Influence of In-Plane Load on the Thermal Buckling and Vibration Characteristics of BD-FG Nonuniform Nanobeams 35
3.1 Overview 35
3.2 Mathematical Formulation 36
3.2.1 Energy Relations 37
3.2.2 Nonlocal Governing Equations of Motion 37
3.3 Solution Procedure 39
3.4 Results and Discussion 40
3.5 Summary 47
References 48

4. Influence of In-Plane Load and Nonlinear Thermal Field on the Vibration Response of Embedded BD-FG Nanobeams 50
4.1 Overview 50
4.2 Mathematical Framework 51
4.2.1 Energy Relations 51
4.2.2 Nonlocal Governing Equations of Motion 52
4.3 Numerical Procedure 54
4.4 Results and Discussion 55
4.5 Summary 59
References 60

5. Influence of Surface Energies on Nonlocal Vibrations of BD-FG Nanobeams 62
5.1 Overview 62
5.2 Mathematical Formulation 63
5.2.1 Material Properties of BD-FG Nanobeam with Surface Layers 63
5.2.2 Strain-Displacement Relations 64
5.2.3 Surface Elasticity Theory 65
5.2.4 Nonlocal Stress–Strain Relations and Stress Resultants and Moments 66
5.2.5 Energy Relations 66
5.2.6 Nonlocal Governing Equations of Motion Based on Timoshenko's Beam Theory 67
5.3 Solution Procedure 69
5.4 Results and Discussion 70
5.5 Summary 74
References 74

6. Effect of Nonuniform Geometry of Nanobeam on Nonlocal and Surface Energies of Nanobeam and Vibration Analysis 78
6.1 Overview 78
6.2 Mathematical Formulation 79
6.2.1 Nonlocal Governing Equation of Motion 79
6.3 Solution Procedure 81
6.4 Results and Discussion 82
6.5 Summary 87
References 87

7. Nonlocal Surface Model of Vibrations and Buckling of BD-FG Nanobeam Embedded with Piezoelectric Layers 89
7.1 Overview 89
7.2 Mathematical Formulation of Vibrations for BD-FGP Nanobeam 90

7.2.1 Nonlocal Theory of Piezoelectricity with Surface Effect....91
7.2.2 Surface Elasticity Theory....92
7.2.3 Eringen's Nonlocal Elasticity Theory....92
7.3 Results and Discussion....95
7.4 Summary....98
References....98

8. Conclusions and Future Scope....101
8.1 Overview....101
8.1.1 Generalized Mathematical Models....101
8.1.2 Stochastic Models and Uncertainty Quantification....101
8.1.3 Multi-Field Modelling of Smart Nanostructures....101
8.1.4 AI-Driven Nonlocal Mechanics and Materials Informatics....102
8.2 Summary....102

Index....103

Preface

Who Should Read This Book and Why

Nonlocal Mechanical Behaviour of Functionally Graded Composite Nanostructures is ideal for researchers, graduate and postgraduate students, and professionals in physics, chemistry, computational science, and materials science. This book is especially valuable for those seeking to harness advanced mathematical modelling and analysis skills to transform the understanding and design of functionally graded (FG) nanostructures. The FG materials are a class of advanced composites, which combine desirable properties—such as strength, toughness, thermal resistance, or wear resistance—in a more optimized way than homogeneous or simple composite materials. This book offers a comprehensive guide integrating the mathematical framework based on nonlocal continuum theories with advanced FG nanostructures, making it especially valuable for those looking to overcome the limitations of traditional modelling methods in handling complex systems at the nanoscale. Whether you're new to the field or seeking to deepen your expertise, this text provides both the theoretical foundations and analysis insights necessary to enhance predictive accuracy and computational efficiency in the simulation of advanced FG nanocomposites.

The rapid advancement of nanotechnology has ushered in an era where the design and manipulation of materials at the atomic and molecular scale are not only possible but critical to innovation across fields such as material, energy, biomedicine, and environmental science. Traditional molecular simulations have played a foundational role in this progress, yet their inherent limitations—particularly high computational cost and time requirement—present significant barriers in the study of increasingly complex nanoscale systems. Further, the complexity of designing, manufacturing, modelling, and predicting the behaviour of FG materials at the nanoscale also introduces new challenges and phenomena (e.g., nonuniform stresses, complex thermal expansion behaviour, and difficulty in standard characterization). Recently, the emergence of a nonlocal mathematical framework as a powerful tool for continuum modelling and analysis has opened new pathways for addressing these challenges. This book aims to provide a comprehensive and accessible introduction to the application of nonlocal and surface theories, combined with numerical simulations, for investigating the mechanical behaviour of FG nanostructures. It brings together foundational concepts, advanced modelling techniques, and analytical methods to navigate the dynamic and rapidly evolving landscape of this interdisciplinary field. Our goal is to equip scientists, engineers, research scholars, and students with the knowledge and tools necessary to accelerate discovery, optimize material properties, and deepen our understanding of nanoscale phenomena through the synergy of computational modelling and intelligent analysis.

Existing Works and the Need for This Book

Over the past decade, remarkable advancements have been achieved in both nonlocal continuum modelling and the design of smart nanocomposites, with each domain playing a pivotal role in driving scientific and technological innovation. Existing literature has explored the individual strengths of these domains—continuum modelling and analysis of smart nanocomposites offer deep physical insights, while nonlocal modelling brings unprecedented speed and predictive accuracy to complex composite nanostructures. A large number of studies have demonstrated the promise of integrating these techniques, particularly in the context of nanomaterials, where understanding behaviour at the atomic scale is essential for innovation. However, current resources that often address this integration in the context of FG nanocomposites are limited. This book seeks to bridge that gap by providing a unified, coherent treatment of the challenges and opportunities in FG nanomaterials research. By offering a structured overview of continuum theories, computational methods, tools, and applications, this book fills a critical need for a comprehensive, interdisciplinary reference that empowers both newcomers and seasoned researchers to fully harness the potential of this transformative approach.

What You Will Find in This Book

This book covers the essential fundamentals, theoretical and numerical aspects, and important references related to different modelling approaches specifically applicable to the numerical simulation of FG nanostructures in the light of nonlocal theories. This book is a structured and in-depth exploration of nonlocal theories that can be effectively integrated with numerical simulation techniques to study, analyse, and optimize FG nanostructures. The content is organized to guide you from foundational concepts to advanced applications. It begins with an overview of the basics of nonlocal continuum theory, followed by a numerical technique to simulate the behaviour of the FG nanostructure, ensuring a common understanding regardless of your background. The subsequent chapters discuss the surface elasticity theory used to enhance the accuracy, efficiency, and scalability of simulations at the nanoscale.

Readers will find the techniques for modelling, numerical simulation, and validation of the results, with a strong emphasis on physical interpretability and domain relevance. Whether you are a researcher, student, or practitioner, this book equips you with the theoretical insight and practical tools needed to drive innovation at the intersection of nanotechnology, computational science, and material engineering.

Chapter 1 provides a general introduction to highlight the motivation for the use of nonlocal continuum theories and to consider functionally graded nanomaterials in engineering. Chapter 2 presents the modelling and thermo-mechanical vibration analysis of the bi-directionally functionally graded (BD-FG) nonuniform nanobeam based on the nonlocal continuum framework. Chapter 3 addresses the influence of in-plane load on the thermal buckling and vibration characteristics of the BD-FG nonuniform nanobeam. Chapter 4 explores the impact of the elastic foundation on the nonlinear thermal vibration response of the embedded BD-FG nanobeam. Chapter 5 introduces a comprehensive surface elasticity-based nonlocal mathematical framework for studying the influence of surface energies on nonlocal vibrations of the BD-FG nanobeam. Chapter 6 reviews the prospects of introducing nonuniformity in the geometry of the BD-FG nanobeam and focusing on its role in the nonlocal and surface energies of the nanobeam. Chapter 7 presents an integrated mathematical framework based on nonlocal and surface theory together with the piezoelectric effect for vibrations and buckling of the BD-FG nanobeam embedded with piezoelectric layers. Finally, Chapter 8 delves into the importance and the future scope of nonlocal continuum theories, surface theories, and piezoelectric awareness in model development. The scope of this book includes a wide array of modelling techniques and computational methods, which are crucial for gaining a comprehensive understanding of functionally graded nanostructures.

In this context, the authors are deeply indebted to numerous colleagues, students, collaborators, and mentors. We are genuinely thankful to all of them for numerous stimulating scientific discussions, exchanges of ideas, and, on many occasions, direct contributions towards the intellectual content of the book. Support and encouragement from the Indian Institute of Technology, Roorkee, such as that from Professor R. Lal, are greatly acknowledged.

Chinika Dangi
Susmita Naskar
October 2025

1

Introduction to Nonlocal Continuum Theory and Composite Nanostructures

1.1 Overview

Nanotechnology has been the subject of interest in the field of science and technology since the discovery of carbon nanotubes (CNTs) in 1991 by Japanese scientist Sumio Iijima (Oregan and Gratzel 1991). The growing advancement in nanotechnology has highlighted the potential applications of nanostructures in highly sensitive nanodevices such as actuators (Selvan et al. 2022), atomic force microscopes (AFMs) (Rahaeifard, Kahrobaiyan, and Ahmadian 2009), thin films in the form of shape memory alloys (Fu, Du, and Zhang 2003), micro-electro-mechanical systems (MEMS) and nano-electro-mechanical systems (NEMS) (Witvrouw and Mehta 2005), and nano-sensors. Nanostructures are stronger, lighter, and more sensitive, with a tunable frequency range. Additionally, the specific characteristic of the large surface-to-volume ratio of nanostructures significantly increases the available surface area for interactions, thereby improving the sensitivity and functionality of these devices (Li, Lin, and Ng 2020; Li, Zhang, and Lee 2014). The integration of nanostructures like element-based nanodevices with the Internet of Things (IoT) and Artificial Intelligence (AI) technologies has introduced a transformative advancement to engineering and real-life applications, see Figure 1.1.

These tremendous applications of nanostructures resulted in a quick growth in both experimental research and theoretical research. At first, vibration responses of nano-scaled structures were studied using experimental techniques and molecular dynamics simulations (Bourchak et al. 2020; Liti et al. 1993; Scarpa, Adhikari, and Srikantha Phani 2009; Garcia-Sanchez et al. 2008). Due to the presence of a large number of atoms and molecules in small-scale structures, the experimental processes and atomistic simulations are time-consuming and quite expensive in comparison to continuum mechanics theories (see Figure 1.2) (Kundu et al. 2023; Li and Chou 2003; Zhang, Pei, and Wang 2012; Yu et al. 2013). In view of this, researchers opted for the classical continuum theory for investigating the small-scaled structures (Govindjee and Sackman 1999; Yoon, Ru, and Mioduchowski 2003). According to classical

DOI: 10.1201/9781003583806-1

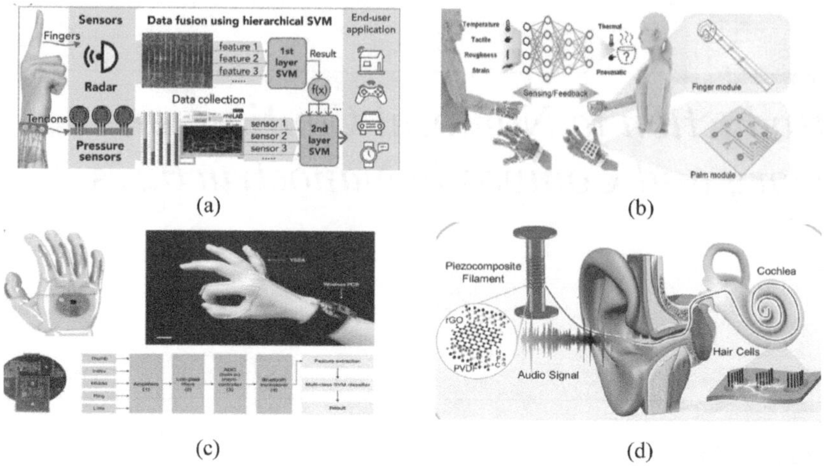

FIGURE 1.1
Applications of nanostructure-based nanodevices integrated with IoT (a) radar and pressure sensors for communication devices (Liang et al. 2019), (b) advanced AI-enhanced wearable glove sensors for sign language recognition (C. Wang et al. 2023), (c) strain sensor-based structure for the smart glove application (Zhou et al. 2020), (d) vibration sensors for communication devices (Malakooti, Zhou, and Sodano 2018; Y. Wang et al. 2023).

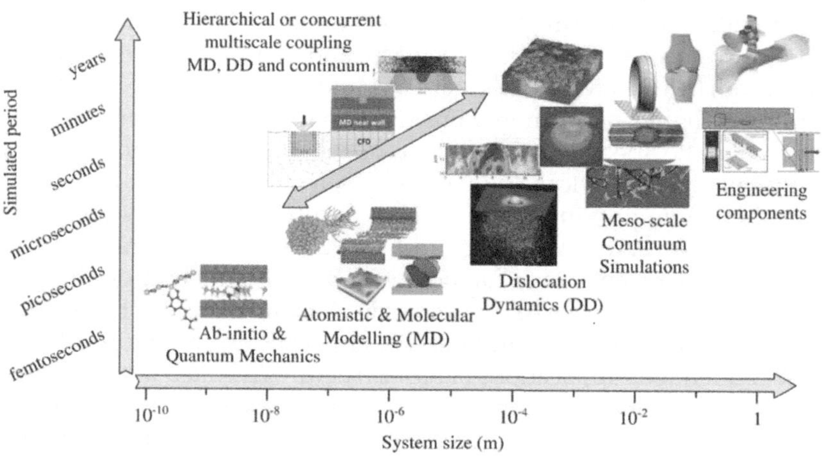

FIGURE 1.2
Time scale vs Length scale representation of various simulation methods.

continuum theory, the behaviour of structures is independent of the effect of intermolecular forces, surface effects, lattice spacing, electric forces, and chemical bonds (Chen, Lee, and Eskandarian 2004; Wang, Tan, and Zhang 2006). However, the experimental results reveal that the small-scale effects can't be neglected at the nanoscale (Hu, Liew, and Wang 2012). Due to this limitation, the use of classical (local) continuum theory is questionable at the small scale (Harik 2001).

Therefore, nonlocal continuum mechanics models came into the picture, initiated by the work of Eringen (Eringen 1983, 2002) and Eringen and Elden (1972). Nonlocal continuum theories take care of small-scale effects by incorporating the size-dependent parameters in the continuum theory-based models. It was Peddieson, Buchanan, and McNitt (2003) who first revealed that the nonlocal continuum theory analyses the behaviour of nanostructures more appropriately. With time, various amendments of nonlocal continuum theories have been proposed in the literature such as strain gradient theory (SGT) (Akgöz and Civalek 2012; Challamel 2013; Mindlin and Eshel 1968; Papargyri-Beskou et al. 2003; Papargyri-Beskou, Giannakopoulos, and Beskos 2010; Tsepoura et al. 2002), couple stress theory (CST) (Wang, Lin, and Zhou 2014; Yang et al. 2002), modified couple stress theory (MCST) (Arefi and Kiani 2017; Ebrahimi and Barati 2018a), nonlocal strain gradient theory (NSGT) (Lim, Zhang, and Reddy 2015; Lu, Guo, and Zhao 2017), and Eringen's nonlocal elasticity theory (ENT) (Civalek and Demir 2016; Ebrahimi, Dehghan, and Seyfi 2019; Lei, Adhikari, and Friswell 2013; Murmu and Adhikari 2010; Murmu, Adhikari, and Wang 2011; Zenkour 2018). In this regard, a few more significant contributions are reported in references (Analooei, Azhari, and Heidarpour 2013; Ansari, Gholami, and Ajori 2013; Chakraverty and Behera 2014; Farajpour et al. 2011; Jena and Chakraverty 2018; Karamooz Ravari and Shahidi 2013; Malikan 2018; Mohammadi, Ghayour, and Farajpour 2013; Murmu and Pradhan 2009; Pradhan and Phadikar 2009; Reddy 2007).

In addition to the above, due to the large surface-to-volume ratio of nanostructures, surface effects cause significant changes in their mechanical behaviour, as reported in the works (Guo and Zhao 2007; Gurtin and Ian Murdoch 1978; Li, Zhang, and Yong 2014), to mention a few. The surface effect is of great importance in surgical devices, drug delivery (Liu et al. 2008), detection of explosives in aviation security (Calzzani et al. 2008), highly sensitive detectors (Malani, Chaudhari, and Sambhe 2015), etc. Theoretically, the difference due to surface effect is approximated using the surface elasticity model described by Gurtin and Murdoch (Gurtin and Ian Murdoch 1975; Murdoch 2005). According to surface elasticity theory (SET), surfaces of the nanomaterials are treated as zero-thickness smooth layers with distinct physical properties from the bulk part. Using Gurtin and Murdoch surface elasticity theory (G-M SET), many researchers have studied the mechanical behaviour of nanostructures, and prominent ones have been reported in references (Eltaher et al. 2013; Farajpour, Dehghany, and Shahidi 2013; Hosseini and Rahmani 2016; Jena et al. 2019, 2020; Lei et al. 2012; Preethi, Rajagopal, and Narasimha 2015; Streitz, Cammarata, and Sieradzki 1994; Wang and Wang 2014).

In the past few decades, there has been an extreme interest in the development of manmade composite materials. The optimization of cost, enhanced sensitivity, excellent flexibility, stretchability, etc., recommends the development of effective functional flexible materials. Among all the composites, functionally graded materials (FGMs) exhibit remarkable properties. The concept of FGMs was initially introduced as a thermal barrier by Japanese

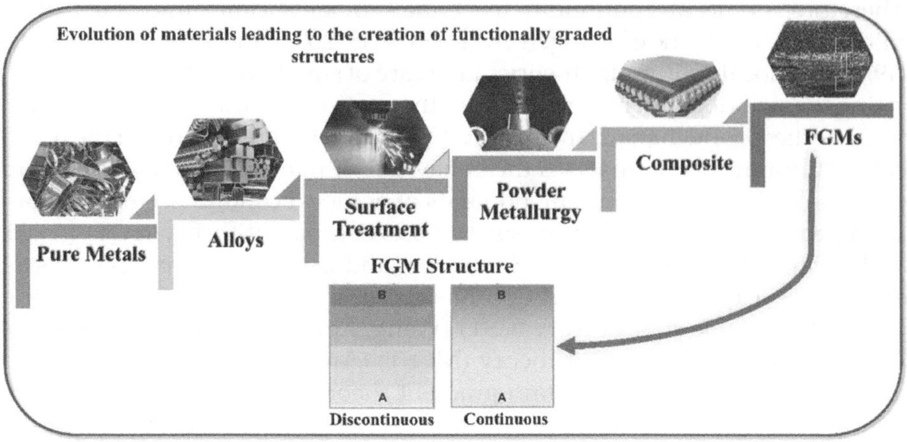

FIGURE 1.3
An evolution in the engineering material from pure metal to the advanced FGMs.

scientists in 1984. Generally, FGMs have been tailored by mixing two or more materials to obtain the combined characteristics of both materials. The most widely used manufacturing techniques are the wet powder coating method, directional solidification method, centrifugal casting method, settlement casting method, filling casting method, powder metallurgy method, vapour deposition method, plasma spraying method, grouting method, etc., which make the change in material properties at the molecular level (Li and Han 2018; Nirmalkumar et al. 2019; Parihar, Setti, and Sahu 2018; Udupa, Rao, and Gangadharan 2014). So far, the material properties of FGMs have been modelled following the distribution functions: (i) Power-law function (Kumari et al. 2017; Singh and Kumari 2018; Susheel et al. 2020), (ii) Mori-Tanaka scheme (Ansari and Norouzzadeh 2016; Belkorissat et al. 2015; Farokhi et al. 2018; Natarajan et al. 2012), (iii) Exponential function (Karami, Janghorban, and Tounsi 2019; Şimşek and Kocatürk 2009). They offer numerous advantages over conventional layered and laminated composites, such as reduced interfacial and thermal stresses, continuous stress distribution, increased bond strength, and adjustable frequency range (see Figure 1.3) (AkhavanAlavi, Mohammadimehr, and Edjtahed 2019; Bharti, Gupta, and Gupta 2013; Sun et al. 2020). The FG nanostructures have demonstrated remarkable potential applications in nanodevices (Hosseini et al. 2018; Van Den Kerkhof et al. 2020; Witvrouw and Mehta 2005).

With these developments, a rapid growth in theoretical and computational studies on FGMs nanostructures has also been noticed. Using nonlocal continuum theories, a large number of studies dealing with the static/dynamic behaviour of FG nanostructures of various geometries under various constraints have appeared in the literature. A complete literature survey of the work on FG nanostructures up to 2019 has been presented by Farajpour et al. (2018). Additionally, some important recent work on FG nanostructures is reported in references (Ahouel et al. 2016; Ansari and Norouzzadeh 2016; Daneshmehr, Rajabpoor, and Hadi 2015; Ebrahimi and Barati 2017, 2018b; Ebrahimi and Salari 2015; Eltaher et al. 2014; Eltaher, Emam, and Mahmoud 2012, 2013; Nazemnezhad and Hosseini-Hashemi 2014; Rahmani and Jandaghian 2015; Rahmani and Pedram 2014).

In the above studies, FGMs are assumed to have a variation of mechanical properties only in one direction. However, in various technological situations, e.g., modern space shuttles (Nemat-alla 2003; Sun et al. 2020), materials have to be operated under thermal or mechanical loads varying in more than one direction. Hence, the conventional unidirectional FGMs may not be appropriate to resist severe multidirectional variations of thermal and mechanical loads. This necessitates developing novel FGMs with material properties varying in two or more directions. A review article compiling the work up to 2020 on numerical modelling and analysis of multidirectional functionally graded structures has been presented by Ghatage, Kar, and Sudhagar (2020). So far, the work related to the static and dynamic behaviour of BD-FG nanobeams and nanoplates is very limited and reported in references (Nejad and Hadi 2016a, 2016b; Nejad, Hadi, and Rastgoo 2016). Beyond the above, only a few studies that cope with thermo-mechanical behaviour of BD-FG nanobeams have appeared in the literature, and significant ones are reported in references (Khaniki and Rajasekaran 2018; Mahinzare, Barooti, and Ghadiri 2018; Mirjavadi, Afshari, et al. 2017; Mirjavadi, Matin, et al. 2017; Shafiei and Kazemi 2017).

1.2 Numerical Techniques

Due to the advancement in high-speed digital computers and computation facilities, a speedy development in computational and numerical techniques has been noticed. Nowadays, researchers have opted for advanced numerical techniques such as Spline interpolation technique (Soni and Sankara Rao 1974), Frobenius method (Ghadiri, Hosseini, and Shafiei 2016), Chebyshev collocation method (Sari, Al-Kouz, and Atieh 2018), finite difference method (FDM) (Karamooz Ravari and Shahidi 2013; Ruocco, Zhang, and Wang 2018), Galerkin's method (Abdullah et al. 2020; Malekzadeh and Farajpour 2012), Rayleigh-Ritz method (Chakraverty and Behera 2014; Pradhan and Chakraverty 2013), meshless method (Roque et al. 2013), differential quadrature method (DQM) (Ebrahimi and Hosseini 2016; Hajhosseini 2020; Khaniki and Hosseini-Hashemi n.d.; Patil and Kadoli 2020; Shu 2011), characteristic orthogonal polynomials (Liew, Hung, and Lim 1995), and differential transform method (DTM) (Ebrahimi et al. 2015; Rahmani et al. 2017), for the static and dynamic characteristics of nanostructures having different geometries with various complicating effects. Among all the methodologies,

DQM, developed by Bellman and Casti (1971), with its various versions: harmonic differential quadrature method (HDQM), differential quadrature element method (DQEM) (Majid Ghadiri, Shafiei, and Alireza Mousavi 2016; Tornabene et al. 2015), generalized differential quadrature method (GDQM) (Sahmani and Safaei 2019), generalized differential quadrature rule (GDQR) (Soltani, Bahar, and Farshidianfar 2011; Wu, Wang, and Liu 2002), etc. are capable of solving these problems without using a large amount of computer memory. A significant number of papers have appeared in the literature showing the capability of this method for producing highly accurate results with minimum computational efforts for both initial and boundary value problems, and recent ones for FG nanobeams and nanoplates are reported in the references (Ebrahimi and Heidari 2019; Mahinzare, Barooti, and Ghadiri 2018; Mahinzare, Ranjbarpur, and Ghadiri 2018; Mirjavadi, Matin, et al. 2017; Mohammadi et al. 2017; Patil and Kadoli 2020; Sahmani and Safaei 2019). On the contrary, the finite element method (FEM), with its various forms (Ansari, Rajabiehfard, and Arash 2010; Do et al. 2017; Kumar et al. 2019; Marakala and Kadoli 2010; Parida and Mohanty 2017; Pradhan and Chakraverty 2015; Susheel et al. 2020; Tornabene et al. 2015; Tornabene, Fantuzzi, and Bacciocchi 2019), is one of the important tools for dealing with complex problems.

1.3 Nonlocal Characteristics of BD-FG Nanostructures

This book aims to provide a framework for mathematical modelling and simulating the vibration and buckling characteristics of smart BD-FG of various nanobeam structures, considering factors such as material composition, geometry, temperature, in-plane forces, and foundation effects. It utilizes classical and first-order shear deformation theories alongside nonlocal continuum theories, deriving governing equations via Hamilton's energy principle.

1.3.1 Organization of the Book

Hereafter, the book is organized as follows:

Chapter 2 introduces the fundamental aspects of modelling the mathematical framework for vibration and buckling of the BD-FG nanobeam in a thermal environment. It clearly outlines the essential components required in formulating and conducting numerical simulations, offering a detailed, step-by-step methodology based on the GDQM. Additionally, it establishes a standardized modelling approach that serves as the basis for the analysis presented in *Chapters 3* and *4*.

Chapters 3 and *4* investigate the effects of in-plane loads and the Pasternak foundation on the vibration and buckling behaviour of BD-FG nonuniform nanobeams under thermal environments, respectively.

Chapter 5 presents an overview of modelling the BD-FG nanobeam with surface layers based on G-M SET integrated with nonlocal theory. This is followed by numerical simulations and analysis of the vibration behaviour, carried out using GDQM.

Chapter 6 extends the analysis from *Chapter 5* by presenting a comprehensive study on the importance of thickness variation on the vibration characteristics of nonuniform BD-FG nanobeams with surface layers.

Chapter 7 presents an integrated mathematical framework that combines nonlocal elasticity theory, surface effects, and the piezoelectric effect to analyse the vibration and buckling behaviour of BD-FG nanobeams embedded with piezoelectric layers.

Chapter 8 discusses current trends and future directions in nonlocal continuum theory, with a particular focus on the incorporation of advanced composite materials in model development.

References

Abdullah, Sardar S., Shahrokh Hosseini-Hashemi, Nazhad A. Hussein, and Reza Nazemnezhad. 2020. "Thermal Stress and Magnetic Effects on Nonlinear Vibration of Nanobeams Embedded in Nonlinear Elastic Medium." *Journal of Thermal Stresses* 43(10):1316–32.

Ahouel, Mama, Mohammed Sid Ahmed Houari, E. A. Adda Bedia, and Abdelouahed Tounsi. 2016. "Size-Dependent Mechanical Behaviour of Functionally Graded Trigonometric Shear Deformable Nanobeams Including Neutral Surface Position Concept." *Steel and Composite Structures* 20(5):963–81.

Akgöz, Bekir, and Ömer Civalek. 2012. "Analysis of Micro-Sized Beams for Various Boundary Conditions Based on the Strain Gradient Elasticity Theory." *Archive of Applied Mechanics* 82(3):423–43.

Akhavan Alavi, Seyed Mohammad, Mehdi Mohammadimehr, and S. H. Edjtahed. 2019. "Active Control of Micro Reddy Beam Integrated with Functionally Graded Nanocomposite Sensor and Actuator Based on Linear Quadratic Regulator Method." *European Journal of Mechanics, A/Solids* 74(October 2018):449–61.

Analooei, Hamid Reza, Mojtaba Azhari, and Amin Heidarpour. 2013. "Elastic Buckling and Vibration Analyses of Orthotropic Nanoplates Using Nonlocal Continuum Mechanics and Spline Finite Strip Method." *Applied Mathematical Modelling* 37(10–11):6703–17.

Ansari, Reza, Raheb Gholami, and Shahram Ajori. 2013. "Torsional Vibration Analysis of Carbon Nanotubes Based on the Strain Gradient Theory and Molecular Dynamic Simulations." *Journal of Vibration and Acoustics, Transactions of the ASME* 135(5):1–6.

Ansari, Raza, and Amir Mohammad Norouzzadeh. 2016. "Nonlocal and Surface Effects on the Buckling Behaviour of Functionally Graded Nanoplates: An Isogeometric Analysis." *Physica E: Low-Dimensional Systems and Nanostructures* 84:84–97.

Ansari, Raza, Reza Rajabiehfard, and Behrouz Arash. 2010. "Nonlocal Finite Element Model for Vibrations of Embedded Multi-Layered Graphene Sheets." *Computational Materials Science* 49(4):831–38.

Arefi, Mohammad, and Masoud Kiani. 2017. "Size-Dependent Free Vibration Analysis of a Three-Layered Exponentially Graded Piezomagnetic Face Sheets Resting on Pasternak's Foundation via MCST." *Journal of Sandwich Structures and Materials* 22(1):1–32.

Belkorissat, Ismahene et al. 2015. "On Vibration Properties of Functionally Graded Nano-Plate Using a New Nonlocal Refined Four Variable Model." *Steel and Composite Structures* 18(4):1063–81.

Bellman, Richard, and John Casti. 1971. "Differential Quadrature and Long-Term Integration." *Journal of Mathematical Analysis and Applications* 34(2):235–38.

Bharti, Isha, Nishu Gupta, and K. M. Gupta. 2013. "Novel Applications of Functionally Graded Nano, Optoelectronic and Thermoelectric Materials." *International Journal of Materials, Mechanics and Manufacturing* 1(3):221–24.

Bourchak, Mostefa, Khalid A. Juhany, Numan Salah, Rafic M. Ajaj, Abeer D. Algarni, and Fabrizio Scarpa. 2020. "Determining the Tensile Properties and Dispersion Characterization of CNTs in Epoxy Using Tem and Raman Spectroscopy." *Mechanics of Composite Materials* 56(2):215–26.

Calzzani, Jr., F. A. et al. 2008. "Detection of Residual Traces of Explosives by Surface Enhanced Raman Scattering Using Gold Coated Substrates Produced by Nanospheres Imprint Technique." *Optics and Photonics in Global Homeland Security IV* 6945(April 2008):694510.

Chakraverty, Snehashish, and Laxmi Behera. 2014. "Free Vibration of Rectangular Nanoplates Using Rayleigh-Ritz Method." *Physica E: Low-Dimensional Systems and Nanostructures* 56:357–63.

Challamel, Noël. 2013. "Variational Formulation of Gradient or/and Nonlocal Higher-Order Shear Elasticity Beams." *Composite Structures* 105:351–68.

Chen, Youping, James D. Lee, and Azim Eskandarian. 2004. "Atomistic Viewpoint of the Applicability of Microcontinuum Theories." *International Journal of Solids and Structures* 41(8):2085–97.

Civalek, Ömer, and Cigdem Demir. 2016. "A Simple Mathematical Model of Microtubules Surrounded by an Elastic Matrix by Nonlocal Finite Element Method." *Applied Mathematics and Computation* 289:335–52.

Daneshmehr, Alireza, Amir Rajabpoor, and Amin Hadi. 2015. "Size Dependent Free Vibration Analysis of Nanoplates Made of Functionally Graded Materials Based on Nonlocal Elasticity Theory with High Order Theories." *International Journal of Engineering Science* 95:23–35.

Do, Thom Van et al. 2017. "Structures Analysis of Bi-Directional Functionally Graded Plates by FEM and a New Third-Order Shear Deformation Plate Theory." *Thin Walled Structures* 119(October):687–99.

Ebrahimi, Farzad et al. 2015. "Application of the Differential Transformation Method for Nonlocal Vibration Analysis of Functionally Graded Nanobeams." *Journal of Mechanical Science and Technology* 29(3):1207–15.

Ebrahimi, Farzad, and Mohammad Reza Barati. 2017. "A Nonlocal Strain Gradient Refined Beam Model for Buckling Analysis of Size-Dependent Shear-Deformable Curved FG Nanobeams." *Composite Structures* 159:174–82.

Ebrahimi, Farzad, and Mohammad Reza Barati. 2018a. "A Modified Nonlocal Couple Stress-Based Beam Model for Vibration Analysis of Higher-Order FG Nanobeams." *Mechanics of Advanced Materials and Structures* 25(13):1121–32.

Ebrahimi, Farzad, and Mohammad Reza Barati. 2018b. "Vibration Analysis of Piezoelectrically Actuated Curved Nanosize FG Beams via a Nonlocal Strain-Electric Field Gradient Theory." *Mechanics of Advanced Materials and Structures* 25(4):350–59.

Ebrahimi, Farzad, M. Dehghan, and Ali Seyfi. 2019. "Eringen's Nonlocal Elasticity Theory for Wave Propagation Analysis of Magneto-Electro-Elastic Nanotubes." *Advances in Nano Research* 1:1–11.

Ebrahimi, Farzad, and Ebrahim Heidari. 2019. "Surface Effects on Nonlinear Vibration of Embedded Functionally Graded Nanoplates via Higher Order Shear Deformation Plate Theory." *Mechanics of Advanced Materials and Structures* 26(8):671–99.

Ebrahimi, Farzad, and Seyed Hamed Seyed Hosseini. 2016. "Thermal Effects on Nonlinear Vibration Behaviour of Viscoelastic Nanosize Plates." *Journal of Thermal Stresses* 39(5):606–25.

Ebrahimi, Farzad, and Erfan Salari. 2015. "A Semi-Analytical Method for Vibrational and Buckling Analysis of Functionally Graded Nanobeams Considering the Physical Neutral Axis Position." *CMES - Computer Modeling in Engineering and Sciences* 105(2):151–81.

Eltaher, Mohamed A., Samir A. Emam, and Fatin Faheem Mahmoud. 2012. "Free Vibration Analysis of Functionally Graded Size-Dependent Nanobeams." *Applied Mathematics and Computation* 218(14):7406–20.

Eltaher, Mohamed A., Samir A. Emam, and Fatin Faheem Mahmoud. 2013. "Static and Stability Analysis of Nonlocal Functionally Graded Nanobeams." *Composite Structures* 96:82–8.

Eltaher, Mohamed A., A. Khairy, Ayman M. Sadoun, and Fatema Alzahraa Omar. 2014. "Static and Buckling Analysis of Functionally Graded Timoshenko Nanobeams." *Applied Mathematics and Computation* 229:283–95.

Eltaher, Mohamed A., Fatin Faheem Mahmoud, Amr E. Assie, and Efsftathios I. Meletis. 2013. "Coupling Effects of Nonlocal and Surface Energy on Vibration Analysis of Nanobeams." *Applied Mathematics and Computation* 224:760–74.

Eringen, A. Cemal. 1983. "On Differential Equations of Nonlocal Elasticity and Solutions of Screw Dislocation and Surface Waves." *Journal of Applied Physics* 54(9):4703–10.

Eringen, A. Cemal. 2002. *Nonlocal Continuum Field Theories*. Springer New York.

Eringen, A. Cemal, and Dominic Gardiner Bowling Edelen. 1972. "On Nonlocal Elasticity." *International Journal of Engineering Science* 10(3):233–48.

Farajpour, Ali, Mohammad Dehghany, and Alireza Shahidi. 2013. "Surface and Nonlocal Effects on the Axisymmetric Buckling of Circular Graphene Sheets in Thermal Environment." *Composites Part B: Engineering* 50:333–43.

Farajpour, Ali, M. Mohammadi, Alireza Shahidi, and Mojtaba H. Mahzoon. 2011. "Axisymmetric Buckling of the Circular Graphene Sheets with the Nonlocal Continuum Plate Model A." *Physica E: Low-Dimensional Systems and Nanostructures* 43(10):1820–25.

Farajpour, A., Ghayesh, M. H. and Farokhi, H. 2018. A review on the mechanics of nanostructures. *International Journal of Engineering Science* 133: 231–263. https://doi.org/10.1016/j.ijengsci.2018.09.006

Farokhi, Hamed, Mergen H. Ghayesh, Alireza Gholipour, and Shahid Hussain. 2018. "Modal Interactions and Energy Transfers in Large-Amplitude Vibrations of Functionally Graded Microcantilevers." *JVC/Journal of Vibration and Control* 24(17):3882–93.

Fu, Yongqing, Hejun Du, and Sam Zhang. 2003. "Functionally Graded TiN/TiNi Shape Memory Alloy Films." *Materials Letters* 57(20):2995–99.

Garcia-Sanchez, A. M. van der Zande, A. San Paulo, B. Lassagne, P. L. McEuen and A. Bachtold. 2008. "Imaging Mechanical Vibrations in Suspended Graphene Sheets." *Nano Letters*, 8(5): 1399–1403.

Ghadiri, Majid, S. H. S. Hosseini, and Navvab Shafiei. 2016. "A Power Series for Vibration of a Rotating Nanobeam with Considering Thermal Effect." *Mechanics of Advanced Materials and Structures* 23(12):1414–20.

Ghadiri, Majid, Navvab Shafiei, and S. Alireza Mousavi. 2016. "Vibration Analysis of a Rotating Functionally Graded Tapered Microbeam Based on the Modified Couple Stress Theory by DQEM." *Applied Physics A: Materials Science and Processing* 122(9):1–14.

Ghatage, Pankaj S., Vishesh R. Kar, and P. Edwin Sudhagar. 2020. "On the Numerical Modelling and Analysis of Multi-Directional Functionally Graded Composite Structures: A Review." *Composite Structures* 236(November 2019):111837.

Govindjee, Sanjay, and Jerome L. Sackman. 1999. "On the Use of Continuum Mechanics to Estimate the Properties of Nanotubes." *Solid State Communications* 110(4):227–30.

Guo, Jian Gang, and Ya Pu Zhao. 2007. "The Size-Dependent Bending Elastic Properties of Nanobeams with Surface Effects." *Nanotechnology* 18(29):295701.

Gurtin, Morton E., and A. Ian Murdoch. 1975. "A Continuum Theory of Elastic Material Surfaces." *Archive for Rational Mechanics and Analysis* 57(4):291–323.

Gurtin, Morton E., and A. Ian Murdoch. 1978. "Surface Stress in Solids." *International Journal of Solids and Structures* 14(6):431–40.

Hajhosseini, Mohammad. 2020. "Analysis of Complete Vibration Bandgaps in a New Periodic Lattice Model Using the Differential Quadrature Method." *JVC/Journal of Vibration and Control* 26(19–20). doi: 10.1177/1077546320902549.

Harik, Vasyl Mivhael. 2001. "Ranges of Applicability for the Continuum Beam Model in the Mechanics of Carbon Nanotubes and Nanorods." *Solid State Communications* 120(7–8):331–35.

Hosseini, Mohammad, Mohammadreza Mofidi, Ali Jamalpoor, and M. Safi Jahanshahi. 2018. "Nanoscale Mass Nanosensor Based on the Vibration Analysis of Embedded Magneto-Electro-Elastic Nanoplate Made of FGMs via Nonlocal Mindlin Plate Theory." *Microsystem Technologies* 24(5):2295–316. doi: 10.1007/s00542-017-3654-8.

Hosseini, Mohammad, and Omid Rahmani. 2016. "Surface Effects on Buckling of Double Nanobeam System Based on Nonlocal Timoshenko Model." *International Journal of Structural Stability and Dynamics* 16(10):1–17.

Hu, Yan-gao, K. M. Liew, and Qing Wang. 2012. "Modeling of Vibrations of Carbon Nanotubes." *Procedia Engineering* 31:343–47.

Ian Murdoch, Alexander 2005. "Some Fundamental Aspects of Surface Modelling." *Journal of Elasticity* 80:33–52.

Jena, Subrat Kumar, Snehashish Chakraverty, Mohammad Malikan, and Francesco Tornabene. 2019. "Stability Analysis of Single-Walled Carbon Nanotubes Embedded in Winkler Foundation Placed in a Thermal Environment Considering the Surface Effect Using a New Refined Beam Theory." *Mechanics Based Design of Structures and Machines* 49(4):581–95.

Jena, Subrat Kumar, Snehashish Chakraverty, Mohammad Malikan, and Francesco Tornabene. 2020. "Effects of Surface Energy and Surface Residual Stresses on Vibro-Thermal Analysis of Chiral, Zigzag, and Armchair Types of SWCNTs Using Refined Beam Theory." *Mechanics Based Design of Structures and Machines* 50(5):1565–79.

Karami, Behrouz, Maziar Janghorban, and Abdelouahed Tounsi. 2019. "On Pre-Stressed Functionally Graded Anisotropic Nanoshell in Magnetic Field." *Journal of the Brazilian Society of Mechanical Sciences and Engineering* 41(11):1–17.

Karamooz Ravari, Mohammad Reza, and Alireza Shahidi. 2013. "Axisymmetric Buckling of the Circular Annular Nanoplates Using Finite Difference Method." *Meccanica* 48(1):135–44.

Khaniki, Hossein Bakhshi, and Sh Hosseini-Hashemi. "Buckling Analysis of Tapered Nanobeams Using Nonlocal Strain Gradient Theory and a Generalized Differential Quadrature Method." *Materials Research Express* 4:056003.

Khaniki, Hossein Bakhshi, and Sundaramoorthy Rajasekaran. 2018. "Mechanical Analysis of Non-Uniform Bi-Directional Functionally Graded Intelligent Micro-Beams Using Modified Couple Stress Theory." *Materials Research Express* 5(5):055703.

Kumar, Ajay et al. 2019. "Experimental and Numerical Comparisons between Finite Element Method, Element-Free Galerkin Method, and Extended Finite Element Method Predicted Stress Intensity Factor and Energy Release Rate of Cortical Bone Considering Anisotropic Bone Modelling." *Proceedings of the Institution of Mechanical Engineers, Part H: Journal of Engineering in Medicine* 233(8):823–38.

Kumari, Poonam, Agyapal Singh, R. K. N. D. Rajapakse, and Santosh Kapuria. 2017. "Three-Dimensional Static Analysis of Levy-Type Functionally Graded Plate with in-Plane Stiffness Variation." *Composite Structures* 168:780–91.

Kundu, Diptiman, Sushanta Ghuku, Susmita Naskar, and Tanmoy Mukhopadhyay. 2023. "Extreme Specific Stiffness through Interactive Cellular Networks in Bi-Level Micro-Topology." *Advanced Engineering Materials* 25:2201407.

Lei, Xiao Wen, Toshiaki Natsuki, Jin Xing Shi, and Qing Ni. 2012. "Surface Effects on the Vibrational Frequency of Double-Walled Carbon Nanotubes Using the Nonlocal Timoshenko Beam Model." *Composites Part B: Engineering* 43(1):64–9.

Lei, Yongun, Sondipon M. Adhikari, and Michael I. Friswell. 2013. "Vibration of Nonlocal Kelvin-Voigt Viscoelastic Damped Timoshenko Beams." *International Journal of Engineering Science* 66–67:1–13.

Li, Chunyu, and Tsu Wei Chou. 2003. "Single-Walled Carbon Nanotubes as Ultrahigh Frequency Nanomechanical Resonators." *Physical Review B - Condensed Matter and Materials Physics* 68(7):073405.

Li, Li, Rongming Lin, and Teng Yong Ng. 2020. "Contribution of Nonlocality to Surface Elasticity." *International Journal of Engineering Science* 152:103311.

Li, Weikai, and Baohong Han. 2018. "Research and Application of Functionally Gradient Materials." *IOP Conference Series: Materials Science and Engineering* 394(2):022065.

Li, Xian Fang, Hua Zhang, and Kang Yong Lee. 2014. "Dependence of Young's Modulus of Nanowires on Surface Effect." *International Journal of Mechanical Sciences* 81:120–25. doi: 10.1016/j.ijmecsci.2014.02.018.

Liang, Xiangpeng et al. 2019. "Fusion of Wearable and Contactless Sensors for Intelligent Gesture Recognition." *Advanced Intelligent Systems* 1(7):1900088.

Liew, Kim Meow, K. C. Hung, and Mongking Lim. 1995. "Vibration of Mindlin Plates Using Boundary Characterstic Orthogonal Polynomials." *Journal of Sound and Vibration* 182:77–90.

Lim, C. W., Ge Zhang, and Junuthula Narasimha Reddy. 2015. "A Higher-Order Nonlocal Elasticity and Strain Gradient Theory and Its Applications in Wave Propagation." *Journal of the Mechanics and Physics of Solids* 78:298–313.

Liti, Hen Nan, Humphery J. maris, Lambert Ben Freund, Kee yong Lee, H. Luhn, Dieter P. Kern. 1993. "Study of Vibrational Modes of Gold Nanostructures by Picosecond Ultrasonics." *Journal of Applied Physics* 73:37–45.

Liu, Zhuang et al. 2008. "Drug Delivery with Carbon Nanotubes for in Vivo Cancer Treatment." *American Association for Cancer Research* 68(16):6652–60.

Lu, Lu, Xingming Guo, and Jianzhong Zhao. 2017. "Size-Dependent Vibration Analysis of Nanobeams Based on the Nonlocal Strain Gradient Theory." *International Journal of Engineering Science* 116:12–24.

Mahinzare, Mohammad, Mohammad Mostafa Barooti, and Majid Ghadiri. 2018. "Vibrational Investigation of the Spinning Bi-Dimensional Functionally Graded (2-FGM) Micro Plate Subjected to Thermal Load in Thermal Environment." *Microsystem Technologies* 24(3):1695–711.

Mahinzare, Mohammad, Hosein Ranjbarpur, and Majid Ghadiri. 2018. "Free Vibration Analysis of a Rotary Smart Two Directional Functionally Graded Piezoelectric Material in Axial Symmetry Circular Nanoplate." *Mechanical Systems and Signal Processing* 100:188–207.

Malakooti, Mohammad H., Zhi Zhou, and Henry A. Sodano. 2018. "Enhanced Energy Harvesting through Nanowire Based Functionally Graded Interfaces." *Nano Energy* 52(May):171–82. doi: 10.1016/j.nanoen.2018.07.051.

Malani, Akshata S., Anagha D. Chaudhari, and Rajeshkumar U. Sambhe. 2015. "A Review on Applications of Nanotechnology in Automotive Industry." *International Journal of Mechanical and Mechatronics Engineering* 10(1):36–40.

Malekzadeh, Parviz, and Ali Farajpour Ouderji. 2012. "Axisymmetric Free and Forced Vibrations of Initially Stressed Circular Nanoplates Embedded in an Elastic Medium." *Acta Mechanica* 223(11):2311–30.

Malikan, Mohammad. 2018. "Buckling Analysis of a Micro Composite Plate with Nano Coating Based on the Modified Couple Stress Theory." *Journal of Applied and Computational Mechanics* 4(1):1–15.

Marakala, Narasimha, and Ravikiran Kadoli. 2010. "Thermally Induced Vibration of a Simply Supported Beam Using Finite Element Method." *International Journal of Engineering Science* 2(12):7874–79.

Mindlin, Raymond D., and N. N. Eshel. 1968. "On First Strain-Gradient Theories in Linear Elasticity." *International Journal of Solids and Structures* 4(1):109–24.

Mirjavadi, Seyed Sajad, Behzad Mohasel Afshari, et al. 2017. "Thermal Vibration of Two-Dimensional Functionally Graded (2D-FG) Porous Timoshenko Nanobeams." *Steel and Composite Structures* 25(4):415–26.

Mirjavadi, Seyed Sajad, Abdolhamid Matin, et al. 2017. "Thermal Buckling Behaviour of Two-Dimensional Imperfect Functionally Graded Microscale-Tapered Porous Beam." *Journal of Thermal Stresses* 40(10):1201–14.

Mohammadi, Kianoosh, Mohammad Mahinzare, Ali Rajabpour, and Majid Ghadiri. 2017. "Comparison of Modeling a Conical Nanotube Resting on the Winkler Elastic Foundation Based on the Modified Couple Stress." *The European Physical Journal Plus* 132:115.

Mohammadi, M., Mostafa Ghayour, and Ali Farajpour. 2013. "Free Transverse Vibration Analysis of Circular and Annular Graphene Sheets with Various Boundary Conditions Using the Nonlocal Continuum Plate Model." *Composites Part B: Engineering* 45(1):32–42.

Murmu, Tony, and Sondipon Adhikari. 2010. "Nonlocal Transverse Vibration of Double-Nanobeam-Systems." *Journal of Applied Physics* 108(8):083514.

Murmu, Tony, Sondipon Adhikari, and Chenguyan Wang. 2011. "Torsional Vibration of Carbon Nanotube-Buckyball Systems Based on Nonlocal Elasticity Theory." *Physica E: Low-Dimensional Systems and Nanostructures* 43(6):1276–80.

Murmu, Tony, and Suresh Chand Pradhan. 2009. "Vibration Analysis of Nano-Single-Layered Graphene Sheets Embedded in Elastic Medium Based on Nonlocal Elasticity Theory." *Journal of Applied Physics* 105(6):064319.Natarajan, Sundararajan, Subhendu Chakraborty, M. Thangavel, Stephane P.A. Bordas, Timon Rabczuk. 2012. "Size-Dependent Free Flexural Vibration Behaviour of Functionally Graded Nanoplates." *Computational Materials Science* 65:74–80.

Nazemnezhad, Reza, and Shahrokh Hosseini-Hashemi. 2014. "Nonlocal Nonlinear Free Vibration of Functionally Graded Nanobeams." *Composite Structures* 110(1):192–9.Nejad, Mohammad Zamani, and Amin Hadi. 2016a. "Eringen's Non-Local Elasticity Theory for Bending Analysis of Bi-Directional Functionally Graded Euler-Bernoulli Nano-Beams." *International Journal of Engineering Science* 106:1–9.

Nejad, Mohammad Zamani, and Amin Hadi. 2016b. "Non-Local Analysis of Free Vibration of Bi-Directional Functionally Graded Euler-Bernoulli Nano-Beams." *International Journal of Engineering Science* 105:1–11.

Nejad, Mohammad Zamani, Amin Hadi, and Abbas Rastgoo. 2016. "Buckling Analysis of Arbitrary Two-Directional Functionally Graded Euler-Bernoulli Nano-Beams Based on Nonlocal Elasticity Theory." *International Journal of Engineering Science* 103:1–10.

Nemat-Alla, Mahmoud. 2003. "Reduction of Thermal Stresses by Developing Two-Dimensional Functionally Graded Materials." *International Journal of Solids and Structures* 40:7339–56.

Nirmalkumar, R., Ritwik Sohgaura, Ravikiran Kadoli, and Sharnappa Joladarashi. 2019. "Powder Metallurgy Process towards Functional Gradation of Al-Al_2O_3 Metal Ceramic Mixture Samples." *IOP Conference Series: Materials Science and Engineering* 624(1):012013.

Oregan, Brian, and Michael Gratzel. 1991. "A low-cost, high-efficiency solar cell based on dye-sensitized colloidal TiO_2 films." *Nature* 353:737–40.

Papargyri-Beskou, S., Antonios E. Giannakopoulos, and Dimitrios E. Beskos. 2010. "Variational Analysis of Gradient Elastic Flexural Plates under Static Loading." *International Journal of Solids and Structures* 47(20):2755–66.

Papargyri-Beskou, S., Katerina G. Tsepoura, Demosthenes Polyzos, and Dimitrios E. Beskos. 2003. "Bending and Stability Analysis of Gradient Elastic Beams." *International Journal of Solids and Structures* 40(2):385–400.

Parida, Smita, and Sukesh Chandra Mohanty. 2017. "Thermoelastic Vibration Analysis of Functionally Graded Skew Plate Using Nonlinear Finite Element Method." *Journal of Thermal Stresses* 40(9):1111–33.

Parihar, Rityuj Singh, Srinivasu Gangi Setti, and Raj Kumar Sahu. 2018. "Recent Advances in the Manufacturing Processes of Functionally Graded Materials: A Review." *IEEE Journal of Selected Topics in Quantum Electronics* 25(2):309–36.

Patil, Mukund A., and Ravikiran Kadoli. 2020. "Differential Quadrature Solution for Vibration Control of Functionally Graded Beams with Terfenol-D Layer." *Applied Mathematical Modelling* 84:137–57.

Peddieson, John, George R. Buchanan, and Richard P. McNitt. 2003. "Application of Nonlocal Continuum Models to Nanotechnology." *International Journal of Engineering Science* 41(3–5):305–12.

Pradhan, Karan Kumar, and Snehashish Chakraverty. 2013. "Free Vibration of Euler and Timoshenko Functionally Graded Beams by Rayleigh-Ritz Method." *Composites Part B: Engineering* 51:175–84.

Pradhan, Karan Kumar, and Snehashish Chakraverty. 2015. "Free Vibration of Functionally Graded Thin Elliptic Plates with Various Edge Supports." *Structural Engineering and Mechanics* 53(2):337–54.

Pradhan, Suresh Chandra, and Jayanta Kumar Phadikar. 2009. "Nonlocal Elasticity Theory for Vibration of Nanoplates." *Journal of Sound and Vibration* 325(1–2):206–23.

Preethi, Kasirajan, Amirtham Rajagopal, and Junuthula Narasimha. 2015. "Surface and Non-Local Effects for Non-Linear Analysis of Timoshenko Beams." *International Journal of Non-Linear Mechanics* 76:100–11.

Rahaeifard, Masoud, Mohammad Hussein Kahrobaiyan, and Mohamad Taghi Ahmadian. 2009. "Sensitivity Analysis of Atomic Force Microscope Cantilever Made of Functionally Graded Materials." *ASME 2009 International Design Engineering Technical Conferences and Computers and Information in Engineering Conference*: 539–44.
Rahmani, Omid, and Ali Akbar Jandaghian. 2015. "Buckling Analysis of Functionally Graded Nanobeams Based on a Nonlocal Third-Order Shear Deformation Theory." *Applied Physics A: Materials Science and Processing* 119(3):1019–32.

Rahmani, Omid, Alireza Mohammadi Niaei, Seyed Amirhosein Hosseini and M. Shoiaei. 2017. "In-Plane Vibration of FG Micro/Nano-Mass Sensor Based on Nonlocal Theory under Various Thermal Loading via Differential Transformation Method." *Superlattices and Microstructures* 101:23–39.

Rahmani, Omid, and Omid Pedram. 2014. "Analysis and Modeling the Size Effect on Vibration of Functionally Graded Nanobeams Based on Nonlocal Timoshenko Beam Theory." *International Journal of Engineering Science* 77:55–70.

Reddy, Junuthula N. 2007. "Nonlocal theories for bending, buckling and vibration of beams." *International Journal of Engineering Science* 45(2–8):288–307.

Roque, Carla, Daniel S. Fidalgo, Antonio J. M. Ferreira and Junuthula N. Reddy. 2013. "A Study of a Microstructure-Dependent Composite Laminated Timoshenko Beam Using a Modified Couple Stress Theory and a Meshless Method." *Composite Structures* 96:532–37.

Ruocco, Eugenio, Hong Zhand and Chien Ming Wang. 2018. "Buckling and Vibration Analysis of Nonlocal Axially Functionally Graded Nanobeams Based on Hencky-Bar Chain Model." *Applied Mathematical Modelling* 63:445–63.

Sahmani, Saeid, and Babak Safaei. 2019. "Nonlocal Strain Gradient Nonlinear Resonance of Bi-Directional Functionally Graded Composite Micro/Nano-Beams under Periodic Soft Excitation." *Thin-Walled Structures* 143(May):106226.

Sari, Ma'en S., Wael G. Al-Kouz, and Anas Atieh. 2018. "Buckling Analysis of Axially Functionally Graded Tapered Nanobeams Resting on Elastic Foundations, Based on Nonlocal Elasticity Theory." *Journal of Mechanical Engineering* 64(12):772–82.

Scarpa, Fabrizio, Sondipon Adhikari and A. Srikantha Phani. 2009. "Effective Elastic Mechanical Properties of Single Layer Graphene Sheets." *Nanotechnology* 20(6):065709.

Selvan, T. Muthamil et al. 2022. "Printable Carbon Nanotube-Liquid Elastomer-Based Multifunctional Adhesive Sensors for Monitoring Physiological Parameters." *ACS Applied Materials and Interfaces* 14(40):45921–33.

Shafiei, Navvab, and Mohammad Kazemi. 2017. "Buckling Analysis on the Bi-Dimensional Functionally Graded Porous Tapered Nano-/Micro-Scale Beams." *Aerospace Science and Technology* 66:1–11.

Shu, Chang. 2011. *Differential Quadrature and Its Application in Engineering Differential Quadrature and Its Application in Engineering*. Springer London.

Şimşek, Mesut, and Turgut Kocatürk. 2009. "Free and Forced Vibration of a Functionally Graded Beam Subjected to a Concentrated Moving Harmonic Load." *Composite Structures* 90(4):465–73.

Singh, Agyapal, and Poonam Kumari. 2018. "Analytical Solution of Functionally Graded Beam Having Longitudinal Stiffness Variation." *International Journal of Computational Methods in Engineering Science and Mechanics* 19(6):390–95.

Soltani, Payam, Payam Bahar and Anoushiravan Farshidianfar. 2011. "An Efficient GDQ Model for Vibration Analysis of a Multiwall Carbon Nanotube on Pasternak Foundation with General Boundary Conditions." *Proceedings of the Institution of Mechanical Engineers, Part C: Journal of Mechanical Engineering Science* 225(7):1730–41.

Soni, Raj Soni, and K. Sankara Rao. 1974. "Vibrations of Non-Uniform Rectangular Plates: A Spline Technique Method of Solution." *Journal of Sound and Vibration* 35(1):35–45.

Streitz, Frederick, R. C. Cammarata, and Karl Sieradzki. 1994. "Surface-Stress Effects on Elastic Properties. I. Thin Metal Films." *Physical Review B* 49(15):10699–706.

Sun, Rujie et al. 2020. "Tunable, Multi-Modal, and Multi-Directional Vibration Energy Harvester Based on Three-Dimensional Architected Metastructures." *Applied Energy* 264(January):114615.

Susheel, Chanderkant K., Anshul Sharma, Rajeev Kumar, and Vishal S. Chauhan. 2020. "Geometrical Nonlinear Characteristics of Functionally Graded Structure Using Functionally Graded Piezoelectric Materials." *Journal of Sandwich Structures and Materials* 22. doi: 10.1177/1099636217752114.

Tornabene, Francesco, Nicholas Fantuzzi, and Michele Bacciocchi. 2019. "Refined Shear Deformation Theories for Laminated Composite Arches and Beams with Variable Thickness: Natural Frequency Analysis." *Engineering Analysis with Boundary Elements* 100(June 2017):24–47.

Tornabene, Francesco, Nicholas Fantuzzi, Francesco Ubertini, and Erasmo Viola. 2015. "Strong Formulation Finite Element Method Based on Differential Quadrature: A Survey." *Applied Mechanics Reviews* 67(2):020801.

Tsepoura, Katerina G., S. Papargyri-Beskou, Demosthenes Polyzos, and Dimitrios E. Beskos. 2002. "Static and Dynamic Analysis of a Gradient-Elastic Bar in Tension." *Archive of Applied Mechanics* 72(6–7):483–97.

Udupa, Gururaja, S. Shrikantha Rao, and K. V. Gangadharan. 2014. "Functionally Graded Composite Materials: An Overview." *Procedia Materials Science* 5:1291–9.

Van Den Kerkhof, S., M. Blommaert, Jan Willem Coenen, and Simon Heuer. 2020. "Investigating the Potential of FGMs through Numerical Minimization of Thermal Stresses." *Physica Scripta* T171(1):014001.

Wang, Chien Ming, Vincent Tan, and Yingyan Zhang. 2006. "Timoshenko Beam Model for Vibration Analysis of Multi-Walled Carbon Nanotubes." *Journal of Sound and Vibration* 294(4):1060–72.

Wang, Chan et al. 2023. "Artificial Intelligence Enhanced Sensors - Enabling Technologies to next-Generation Healthcare and Biomedical Platform." *Bioelectronic Medicine* 9(1):1–34. doi: 10.1186/s42234-023-00118-1.

Wang, Dong-Hui, and Gang-Feng Wang. 2014. "Surface Effects on the Vibration and Buckling of Double-Nanobeam-Systems Surface Effects on the Vibration and Buckling of Double-Nanobeam-Systems." *Journal of Nanomaterials*. doi: 10.1155/2011/518706.

Wang, Yong Gang, Wen Hui Lin, and Chang Ling Zhou. 2014. "Nonlinear Bending of Size-Dependent Circular Microplates Based on the Modified Couple Stress Theory." *Archive of Applied Mechanics* 84(3):391–400.

Wang, Yuan et al. 2023. "Expedient Secondary Functions of Flexible Piezoelectrics for Biomedical Energy Harvesting." *Bioactive Materials* 22(August 2022):291–311. doi: 10.1016/j.bioactmat.2022.10.003.

Witvrouw, Ann, and Anshu Mehta. 2005. "The Use of Functionally Graded Poly-SiGe Layers for MEMS Applications." *Materials Science Forum* 492–493:255–60.

Wu, Tianyun, Yuyong Wang, and Guirong Liu. 2002. "Free Vibration Analysis of Circular Plates Using Generalized Differential Quadrature Rule." *Computer Methods in Applied Mechanics and Engineering* 191:5365–80.

Yang, Feng, Arthur C. M. Chong, David Chuen Chun Lam, and Pin Tong. 2002. "Couple Stress Based Strain Gradient Theory for Elasticity." *International Journal of Solids and Structures* 39(10):2731–43.

Yoon, J., Chongqing Ru, and Andrzej Mioduchowski. 2003. "Vibration of an Embedded Multiwall Carbon Nanotube." *Composites Science and Technology* 63(11):1533–42.

Yu, Hong, C. Sun, Weiguang Zhang, S. Y. Lei, Qing-An Huang. 2013. "Study on Size-Dependent Young's Modulus of a Silicon Nanobeam by Molecular Dynamics Simulation." *Journal of Nanomaterials*. doi: 10.1155/2013/319302.

Zenkour, Ashraf M. 2018. "Nonlocal Elasticity and Shear Deformation Effects on Thermal Buckling of a CNT Embedded in a Viscoelastic Medium." *The European Physical Journal Plus* 133:196.

Zhang, Yingyan, Qing-Xiang Pei, and Chien Ming Wang. 2012. "Mechanical Properties of Graphynes under Tension: A Molecular Dynamics Study." *Applied Physics Letters* 101:081909.

Zhou, Zhihao et al. 2020. "Sign-to-Speech Translation Using Machine-Learning-Assisted Stretchable Sensor Arrays." *Nature Electronics* 3(9): 571–78. doi: 10.1038/s41928-020-0428-6.

2

Thermo-Mechanical Model of the Vibration Behaviour of BD-FG Nonuniform Nanobeams

2.1 Overview

Thermo-mechanical behaviour of structures is crucial for applications in automotive, aerospace, and water transportation industries, where materials face extreme, multi-directional temperature gradients (Kumar, Mishra, and Jain 2008; Mathew, Joy, and George 2019; Sun et al. 2020). It necessitates the development of advanced FGM structures whose material properties vary continuously not only in one specified direction but also in other directions (Nemat-Alla 2003). In this context, the BD-FG nanostructures are expected to have reduced thermal stresses together with self-monitoring and self-controlling capabilities. In spite of the vast literature on static/dynamic behaviour of one-directional FG nanobeams, very limited work is available on BD-FG nanobeams (Nejad and Hadi 2016; Mirjavadi et al. 2017a, 2017b; Shafiei et al. 2017; Shafiei and Kazemi 2017; Khaniki and Rajasekaran 2018; Mahinzare, Barooti, and Ghadiri 2018; Trinh et al. 2018; Sahmani and Safaei 2019; Tang Lv, and Yang 2019), so far. Among these, only selective articles cope with the buckling and vibration characteristics of BD-FG nanobeams under the influence of a thermal field, considering nonuniformity in the cross-section. Recently, the effect of a nonlinear thermal field on the vibrations of a BD-FG porous nanobeam has been studied by Mirjavadi et al. (2017a) using TBT and Eringen's nonlocal theory. The variation of material properties in the thickness and length direction is assumed to follow a power-law distribution. The numerical solution is obtained using GDQM. Mirjavadi et al. (2017b) explored the thermal buckling behaviour of BD-FG porous tapered nanobeam using MCST and assuming temperature-dependent material properties graded in thickness and width directions according to a power-law model. The buckling behaviour of BD-FG nonuniform nanobeam with variation in mechanical properties according to the power-law

DOI: 10.1201/9781003583806-2

function in both the thickness and length directions has been investigated by Khaniki and Rajasekaran (2018) on the basis of MCST with EBT.

In this chapter, the vibration behaviour of BD-FG nonuniform Timoshenko nanobeam under the influence of thermal field is investigated using first-order shear deformation theory (FSDT) together with ENT. The mechanical properties of the beam are assumed to be temperature-dependent, and a nonlinear temperature distribution (NTD) is applied across the thickness. The material composition of the beam varies according to a power-law distribution in the thickness direction and as an exponential function in the axial direction. Nonuniformity in the cross-section of the beam is assumed due to the linear variation in thickness and width along the length. The governing differential equations of motion for such a beam model have been obtained using Hamilton's energy principle and numerically solved for different combinations of clamped, simply supported, and free boundary conditions, namely, clamped-clamped (C-C), clamped-simply supported (C-S), and simply supported-simply supported (S-S), employing GDQM. Numerical simulation has been performed using the software MATLAB to study the effect of the axially functionally gradient index (AFG), transverse functionally gradient index (TFG), nonlocal parameter, nonuniformity parameters, and temperature difference on the vibration characteristics of the nanobeam for all boundary conditions and the first three modes of vibration, and the results have been validated with published work.

2.2 Mathematical Framework

Referred to the Cartesian coordinate system (x, y, z), consider a BD-FG nonuniform nanobeam of length L, width $b(x)$ and thickness $h(x)$along x, y, z-axes, respectively, subjected to a nonlinearly varying temperature field $T = T(z)$ in the thickness direction z. The middle plane being $z = 0$ with the origin at the end $x = 0$. The top $(z = +h/2)$ and bottom $(z = -h/2)$ surfaces of the beam are fully ceramic and metallic, respectively. The thickness and width of the nanobeam are assumed to vary linearly along the length as $h(x) = h_0\left(1 - \beta_h \frac{x}{L}\right)$ and $b(x) = b_0\left(1 - \beta_b \frac{x}{L}\right)$, respectively, where β_h and β_b denote the rates of the cross-sectional change along the thickness and width directions, respectively. The symbols h_0 and b_0 represent the thickness and width of the beam at the origin (see Figure 2.1).

As per the FSDT (Reddy, 2008), the displacement components at any point of the nanobeam are

$$u_x(x, z, t) = u_0(x, t) + z\psi_x(x, t), \quad u_y = 0, \quad u_z(x, t) = w_0(x, t) \tag{2.2.1}$$

where (u_0, w_0) are axial and transverse displacements of the mid-plane, ψ_x is the rotational angle of the yz-plane about the transverse direction.

The strain components are

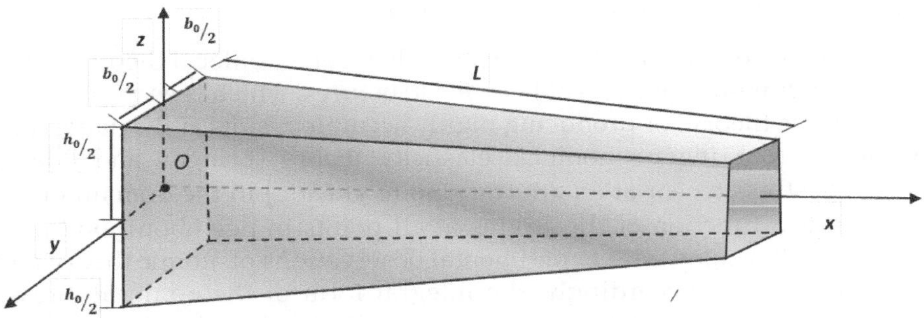

FIGURE 2.1
Geometry of the BD-FG nonuniform nanobeam in Cartesian coordinate system.

$$\varepsilon_{xx} = \frac{\partial u_x}{\partial x} = \frac{\partial u_0}{\partial x} + z\frac{\partial \psi_x}{\partial x}, \quad \gamma_{xz} = \frac{\partial u_x}{\partial z} + \frac{\partial u_z}{\partial x} = \frac{\partial w_0}{\partial x} + \psi_x \tag{2.2.2}$$

2.2.1 Temperature-Dependent Material Properties of BD-FG Nanobeam

Recent developments in the aerospace industry demand functionally graded (FG) nanostructures to enhance the thermal resistance and durability of components such as coatings, engines, drives, and tires exposed to extreme heat (Mathew et al. 2019). Following (Nejad and Hadi 2016), the effective temperature-dependent material property of the BD-FG nanobeam is taken to be graded exponentially in the axial direction and according to a power-law function in the thickness direction. Accordingly, the specific material property P_f of the BD-FG nanobeam at any point is given by,

$$P_f(x,z,T) = e^{n_x(x/L)}\left[\left(P_c(T) - P_m(T)\right)\left(\frac{z}{h} + \frac{1}{2}\right)^{n_z} + P_m(T)\right] \tag{2.2.3}$$

where superscripts n_x and n_z are used for the axially functionally graded index (AFG) and transverse functionally graded index (TFG), respectively. The material property P_f represents the Young's modulus E, shear modulus G, mass density ρ, thermal expansion coefficients α and thermal conductivity k for the BD-FG nanobeam.

The mechanical properties P_c and P_m for beam constituents are taken as follows (Shahrjerdi et al. 2011):

$$P_b(T) = P_0\left(P_{-1}T^{-1} + T^0 + P_1T^1 + P_2T^2 + P_3T^3\right), \quad b = m, c \tag{2.2.4}$$

where $P_0, P_{-1}, P_1, P_2, P_3$ are temperature-dependent coefficients of material properties.

2.2.2 Eringen's Nonlocal Elasticity Theory

A vast literature is available on the use of Eringen's nonlocal theory for the static and dynamic analysis of FG nanostructures, which reveals the capability of this theory for producing highly accurate results at the nanoscale. According to Eringen's nonlocal elasticity theory (Eringen and Edelen 1972), the stress tensor at an arbitrary point $\mathbf{x}(x,y,z)$ in the domain of the material is a function of the strains at all points in neighbourhood of $\mathbf{x}$, which is in harmony with experimental observations of atomic theory and lattice dynamics. Accordingly, the integral form of the nonlocal stress–strain relationship at the reference point $\mathbf{x}$ in the domain of material is expressed as:

$$\sigma_{ij}^{nl} = \int_v \zeta\left(|\mathbf{x}' - \mathbf{x}|, \tau\right) t_{ij}(\mathbf{x}') dv(\mathbf{x}') \tag{2.2.5}$$

where $\zeta\left(|x' - x|, \tau\right)$ is the nonlocal kernel function, influenced by the internal characteristic length, x' is any arbitrary point in the neighbourhood of $\mathbf{x}$, and dv is the elementary volume. The symbol $\tau = e_0 l / L$ represents a small-scale constant, where e_0 is a material constant, l and L are the internal and external characteristic lengths of the nanostructures, respectively. σ_{ij}^{nl} and t_{ij} are the components of nonlocal and conventional stress tensors $\boldsymbol{\sigma}^{\mathrm{nl}}$ and $\mathbf{t}$, respectively.

Classical Stress tensor $\mathbf{t}$ is defined as:

$$\mathbf{t} = \mathbf{C} : \boldsymbol{\varepsilon} \tag{2.2.6}$$

where $\mathbf{C}$ is the fourth-order elasticity tensor and ε is the strain tensor.

Using a physically admissible kernel ζ's, the integral form in equation (2.2.5) reduces to a simplified differential form of the nonlocal stress–strain relationship and given by (Eringen, 1983),

$$\left(1 - \mu \nabla^2\right) \boldsymbol{\sigma}^{\mathrm{nl}} = \mathbf{t} = \mathbf{C} : \boldsymbol{\varepsilon} \tag{2.2.7}$$

where $\mu = (e_0 l)^2$ and known as the nonlocal parameter, revealing the size effect on the response of nanostructures and ∇^2 is the Laplacian operator in Cartesian coordinate systems.

For BD-FG Timoshenko nanobeams, the nonlocal stress–strain equation (2.2.7) is related as follows:

$$\sigma_{xx}^{nl} - \mu \frac{\partial^2 \sigma_{xx}^{nl}}{\partial x^2} = E(x,z,T)\varepsilon_{xx}, \quad \sigma_{xz}^{nl} - \mu \frac{\partial^2 \sigma_{xz}^{nl}}{\partial x^2} = k_s G(x,z,T)\gamma_{xz} \tag{2.2.8}$$

Now, using equation (2.2.2), the nonlocal stress resultants and moments for the BD-FG Timoshenko nanobeam in terms of displacements are:

$$\left.\begin{aligned} N_{xx}^{nl} - \mu \frac{\partial^2}{\partial x^2} N_{xx}^{nl} &= A_{xx} \frac{\partial u_0}{\partial x} + B_{xx} \frac{\partial \psi_x}{\partial x}, \\ M_{xx}^{nl} - \mu \frac{\partial^2}{\partial x^2} M_{xx}^{nl} &= B_{xx} \frac{\partial u_0}{\partial x} + D_{xx} \frac{\partial \psi_x}{\partial x}, \\ Q_{xz}^{nl} - \mu \frac{\partial^2}{\partial x^2} Q_{xz}^{nl} &= C_{xz} \left(\frac{\partial w_0}{\partial x} + \psi_x \right) \end{aligned}\right\} \tag{2.2.9}$$

where $A_{xx}(x), B_{xx}(x), D_{xx}(x)$ are flexural rigidities of the BD-FG nanobeam and given by:

$$\left(A_{xx}(x), B_{xx}(x), D_{xx}(x)\right) = \int_{-b/2}^{b/2} \int_{-h/2}^{h/2} \left(1, z, z^2\right) E(x, z, T)\, dydz \quad \text{and}$$

$$C_{xz}(x) = \int_{-b/2}^{b/2} \int_{-h/2}^{h/2} k_s G(x, z, T)\, dydz$$

2.2.3 Energy Relations

The first variation of strain energy stored in the BD-FG nanobeam is given by:

$$\delta U = \frac{1}{2} \int_0^L b \left(N_{xx}^{nl} \frac{\partial(\delta u_0)}{\partial x} + M_{xx}^{nl} \frac{\partial(\delta \psi_x)}{\partial x} + Q_x^{nl} \left(\frac{\partial(\delta w_0)}{\partial x} + (\delta \psi_x) \right) \right) dx \tag{2.2.10}$$

The variation of energy stored in the BD-FG nanobeam due to the temperature change across the thickness is

$$\delta U^T = \int_0^L b \left(N^T \frac{\partial w_0}{\partial x} \frac{\partial(\delta w_0)}{\partial x} \right) dx \tag{2.2.11}$$

where $N^T(x) = \int_{-h/2}^{h/2} -bE(x, z, T)\alpha(x, z, T)\left(T(z) - T_0\right) dz, T_0$ is the initial temperature.

The first variation of kinetic energy of BD-FG nanobeam is obtained as follows:

$$\delta K=\int_{0}^{L} b\left[\begin{array}{l} I_0\left(\frac{\partial u_0}{\partial t}\frac{\partial(\delta u_0)}{\partial t}+\frac{\partial w_0}{\partial t}\frac{\partial \delta(w_0)}{\partial t}\right)+I_1\left(\frac{\partial \psi_x}{\partial t}\frac{\partial(\delta u_0)}{\partial t}+\frac{\partial u_0}{\partial t}\frac{\partial(\delta \psi_x)}{\partial t}\right) \\ +I_2\left(\frac{\partial \psi_x}{\partial t}\frac{\partial(\delta \psi_x)}{\partial t}\right) \end{array}\right] dx \tag{2.2.12}$$

where $\left(I_0(x),I_1(x),I_2(x)\right)=\int_{-b/2}^{b/2}\int_{-h/2}^{h/2}\left(1,z,z^2\right)\rho(x,z,T)\,dydz.$

2.2.4 Nonlocal Governing Equations of Motion

Using the expressions for energies from equations (2.2.10) to (2.2.12) in the Hamilton's energy principle for $\delta L=\delta K-\delta U-\delta U^T$, the Euler-Lagrange equations are obtained as:

$$\frac{\partial N_{xx}^{nl}}{\partial x}=I_0\frac{\partial^2 u_0}{\partial t^2}+I_1\frac{\partial^2 \psi_x}{\partial t^2} \tag{2.2.13}$$

$$\frac{\partial M_{xx}^{nl}}{\partial x}-Q_{xz}^{nl}=I_1\frac{\partial^2 u_0}{\partial t^2}+I_2\frac{\partial^2 \psi_x}{\partial t^2} \tag{2.2.14}$$

$$\frac{\partial Q_{xz}^{nl}}{\partial x}+N^T\frac{\partial^2 w_0}{\partial x^2}=I_0\frac{\partial^2 w_0}{\partial t^2} \tag{2.2.15}$$

Substituting the expression for nonlocal stresses and moments (2.2.9), the nonlocal governing equations of motion (2.2.13)–(2.2.15) for the considered beam are given by:

$$A_{xx}\frac{\partial^2 u_0}{\partial x^2}+B_{xx}\frac{\partial^2 \psi_x}{\partial x^2}+\frac{\partial A_{xx}}{\partial x}\frac{\partial u_0}{\partial x}+\frac{\partial B_{xx}}{\partial x}\frac{\partial \psi_x}{\partial x}=\left(1-\mu\frac{\partial^2}{\partial x^2}\right)\left(I_0\frac{\partial^2 u_0}{\partial t^2}+I_1\frac{\partial^2 \psi_x}{\partial t^2}\right) \tag{2.2.16}$$

$$\begin{aligned} &B_{xx}\frac{\partial^2 u_0}{\partial x^2}+D_{xx}\frac{\partial^2 \psi_x}{\partial x^2}+\frac{\partial B_{xx}}{\partial x}\frac{\partial u_0}{\partial x}+\frac{\partial D_{xx}}{\partial x}\frac{\partial \psi_x}{\partial x}-C_{xz}\left(\frac{\partial w_0}{\partial x}+\psi_x\right) \\ &=\left(1-\mu\frac{\partial^2}{\partial x^2}\right)\left(I_1\frac{\partial^2 u_0}{\partial t^2}+I_2\frac{\partial^2 \psi_x}{\partial t^2}\right) \end{aligned} \tag{2.2.17}$$

$$C_{xz}\frac{\partial^2 w_0}{\partial x^2}+C_{xz}\frac{\partial \psi_x}{\partial x}+\frac{\partial C_{xz}}{\partial x}\left(\frac{\partial w_0}{\partial x}+\psi_x\right)+\left(1-\mu\frac{\partial^2}{\partial x^2}\right)\frac{\partial}{\partial x}\left(N^T(x)\frac{\partial w_0}{\partial x}\right)$$
$$=\left(1-\mu\frac{\partial^2}{\partial x^2}\right)\left(I_0\frac{\partial^2 w_0}{\partial t^2}\right) \tag{2.2.18}$$

For harmonic vibration, we assume $u_0(x,t)=\tilde{u}_0 e^{i\omega t}$, $w_0(x,t)=\tilde{w}_0 e^{i\omega t}$, $\psi_x(x,t)=\tilde{\psi}_x e^{i\omega t}$ and using non-dimensional variables $\overline{w}_0=\frac{\tilde{w}_0}{L}$, $\overline{u}_0=\frac{\tilde{u}_0}{L}$, $\overline{\psi}_x=\tilde{\psi}_x$, $X=\frac{x}{L}$, $H=\frac{h_0}{L}$ and $Z=\frac{z}{h}$, equations (2.2.16)–(2.2.18) become:

$$U_0\frac{d^2\overline{u}_0}{dX^2}+U_1\frac{d^2\overline{\psi}_x}{dX^2}+\frac{dU_0}{dX}\frac{d\overline{u}_0}{dX}+\frac{dU_1}{dX}\frac{d\overline{\psi}_x}{dX}=-\hat{\Omega}^2\left(1-\gamma^2\frac{d^2}{dX^2}\right)\left[V_0\overline{u}_0+V_1\overline{\psi}_x\right] \tag{2.2.19}$$

$$U_1\frac{d^2\overline{u}_0}{dX^2}+U_2\frac{d^2\overline{\psi}_x}{dX^2}+\frac{dU_1}{dX}\frac{d\overline{u}_0}{dX}+\frac{dU_2}{dX}\frac{d\overline{\psi}_x}{dX}-U_3\left(\frac{d\overline{w}_0}{dX}+\overline{\psi}_x\right)$$
$$=-\hat{\Omega}^2\left(1-\gamma^2\frac{d^2}{dX^2}\right)\left[V_1\overline{u}_0+V_2\overline{\psi}_x\right] \tag{2.2.20}$$

$$U_3\frac{d^2\overline{w}_0}{dX^2}+U_3\frac{d\overline{\psi}_x}{dX}+\frac{dU_3}{dX}\left(\frac{d\overline{w}_0}{dX}+\overline{\psi}_x\right)+\left(1-\gamma^2\frac{d^2}{dX^2}\right)\left(\frac{d}{dX}\left(P_1\frac{d\overline{w}_0}{dX}\right)\right)$$
$$=-\hat{\Omega}^2\left(1-\gamma^2\frac{d^2}{dX^2}\right)\left[V_0\overline{w}_0\right] \tag{2.2.21}$$

together with the boundary conditions as follows:

i. Clamped-Clamped beam:

$$\overline{u}_0=\overline{w}_0=\overline{\psi}_x=0 \quad \text{at} \quad X=0 \quad \text{and} \quad 1 \tag{2.2.22}$$

ii. Clamped-Simply supported boundary beam:

$$\overline{u}_0=\overline{w}_0=\overline{\psi}_x=0 \quad \text{at} \quad X=0 \quad \text{and} \quad \overline{u}_0=\overline{w}_0=M_{xx}^{nl}=0 \quad \text{at} \quad X=1 \tag{2.2.23}$$

iii. Simply supported-Simply supported beam:

$$\bar{u}_0 = \bar{w}_0 = M_{xx}^{nl} = 0 \quad \text{at} \quad X = 0 \quad \text{and} \quad 1 \tag{2.2.24}$$

where

$$U_0 = e^{n_x X}(1-\beta_b X)(1-\beta_h X)\bar{A}_{xx} H, \quad U_1 = e^{n_x X}(1-\beta_b X)(1-\beta_h X)^2 \bar{B}_{xx} H^2$$

$$U_2 = e^{n_x X}(1-\beta_b X)(1-\beta_h X)^3 \bar{D}_{xx} H^3, \quad U_3 = e^{n_x X}(1-\beta_b X)(1-\beta_h X)\bar{C}_{xz} H$$

$$V_0 = e^{n_x X} H (1-\beta_b X)(1-\beta_h X)\bar{I}_0, \quad V_1 = e^{n_x X}(1-\beta_b X)(1-\beta_h X)^2 \bar{I}_1 H^2$$

$$V_2 = e^{n_x X}(1-\beta_b X)(1-\beta_h X)^3 \bar{I}_3 H^3, \quad P_1 = e^{2n_x X}(1-\beta_b X)(1-\beta_h X) H \alpha_0^c \bar{N}^T$$

$$\hat{\Omega}^2 = \frac{\omega^2 L^2 \rho_0^c}{E_0^c}, \bar{N}^T = \frac{N^T}{bhE_0^c}, \left(\bar{A}_{xx}, \bar{B}_{xx}\bar{D}_{xx}\right) = \left(\frac{A_{xx}}{bhE_0^c}, \frac{B_{xx}}{bh^2E_0^c}, \frac{D_{xx}}{bh^3E_0^c}\right), \bar{C}_{xz} = \frac{C_{xz}}{bhE_0^c}, \text{ and}$$
$$\left(\bar{I}_0, \bar{I}_1, \bar{I}_2\right) = \left(\frac{I_0}{bh\rho_0^c}, \frac{I_1}{bh^2\rho_0^c}, \frac{I_2}{bh^3\rho_0^c}\right).$$

The system of coupled differential equations (2.2.19)–(2.2.21) with variable coefficients has been solved using GDQM for appropriate boundary conditions.

2.3 Method of Solution

Literature survey reveals that the GDQM is less computationally expensive and provides a higher rate of convergence than the other numerical schemes (Chakraverty and Behera 2014; Norouzzadeh and Ansari 2017; Rajasekaran and Khaniki 2018). Let X_i, $i = 1, 2, \ldots, n$ be the n grid points in the applicability domain [0, 1] of the beam. According to GDQM (Shu 2011), the rth derivative of a function f at a grid point X_i is approximated as:

$$\frac{d^r f(X_i)}{dX^r} = \sum_{j=1}^{j=n} C_{ij}^{(r)} f(X_j) \tag{2.3.1}$$

where $C_{ij}^{(r)}$'s are weighting coefficients.
0

$$C_{ij}^{(1)} = -\sum_{j=1,\, i\neq j}^{j=n} C_{ij}^{(1)}, \quad i = j$$

$$M(X_i) = \prod_{j=1,\, i\neq j}^{j=n} (X_i - X_j)$$

The higher order weighting coefficients i.e., $r > 1$ are obtained by using the recurrence relation,

$$C_{ij}^{(r)} = \sum_{k=1}^{k=n} C_{ik}^{(r-1)} C_{kj}^{(1)}, \quad i, j = 2, \ldots, n \tag{2.3.2}$$

The grid points are chosen as the zeros of the Gauss–Chebyshev–Lobatto polynomial of order n and given by

$$X_i = \frac{1 - \cos[(i-1)\pi/(n-1)]}{2}, \; i = 1, \ldots, n \tag{2.3.3}$$

Using GDQM to discretize the governing equations (2.2.19)–(2.2.21) at the grid point$X_i\,(i = 2, \ldots, n-1)$ and boundary condition for C-C beam (2.2.22), a complete set of $3n$ equations in $3n$ unknowns is obtained, which can be written in matrix form as

$$\begin{bmatrix} [\mathbf{K}_{dd}] & [\mathbf{K}_{db}] \\ [\mathbf{K}_{bd}^{CC}] & [\mathbf{K}_{bb}^{CC}] \end{bmatrix} \begin{Bmatrix} \{\mathbf{F}_d\} \\ \{\mathbf{F}_b\} \end{Bmatrix} - \hat{\Omega}^2 \begin{bmatrix} [\mathbf{M}_{dd}] & [\mathbf{M}_{db}] \\ [\mathbf{M}_{bd}^{CC}] & [\mathbf{M}_{bb}^{CC}] \end{bmatrix} \begin{Bmatrix} \{\mathbf{F}_d\} \\ \{\mathbf{F}_b\} \end{Bmatrix} = [0] \tag{2.3.4}$$

where $[\mathbf{K}_{ij}]$ **and** $[\mathbf{M}_{ij}]$, $i, j = b, d$ are stiffness and mass matrices, and $\begin{Bmatrix} \{\mathbf{F}_d\} \\ \{\mathbf{F}_b\} \end{Bmatrix}$ is the solution vector.

Similarly, for C-S, S-S and boundary conditions, the governing equation can be written as

$$\begin{bmatrix} [\mathbf{K}_{dd}] & [\mathbf{K}_{db}] \\ [\mathbf{K}_{bd}^{CS}] & [\mathbf{K}_{bb}^{CS}] \end{bmatrix} \begin{Bmatrix} \{\mathbf{F}_d\} \\ \{\mathbf{F}_b\} \end{Bmatrix} - \hat{\Omega}^2 \begin{bmatrix} [\mathbf{M}_{dd}] & [\mathbf{M}_{db}] \\ [\mathbf{M}_{bd}^{CS}] & [\mathbf{M}_{bb}^{CS}] \end{bmatrix} \begin{Bmatrix} \{\mathbf{F}_d\} \\ \{\mathbf{F}_b\} \end{Bmatrix} = [0] \tag{2.3.5}$$

$$\begin{bmatrix} [\mathbf{K}_{dd}] & [\mathbf{K}_{db}] \\ [\mathbf{K}_{bd}^{SS}] & [\mathbf{K}_{bb}^{SS}] \end{bmatrix} \begin{Bmatrix} \{\mathbf{F}_d\} \\ \{\mathbf{F}_b\} \end{Bmatrix} - \hat{\Omega}^2 \begin{bmatrix} [\mathbf{M}_{dd}] & [\mathbf{M}_{db}] \\ [\mathbf{M}_{bd}^{SS}] & [\mathbf{M}_{bb}^{SS}] \end{bmatrix} \begin{Bmatrix} \{\mathbf{F}_d\} \\ \{\mathbf{F}_b\} \end{Bmatrix} = [0] \tag{2.3.6}$$

2.4 Results and Discussion

By developing the MATLAB code, the lowest three eigenvalues of equations (2.3.4)–(2.3.6) are retained as the frequency parameter for the first three modes of vibration. For convenience, the non-dimensional frequency parameter $\hat{\Omega}$ has been transformed to Ω using the relation $\Omega = \left(\frac{\sqrt{12}L}{h_0}\right)\hat{\Omega}$. From the literature, temperature-dependent material properties of constituent materials are given in

Table 2.1. The values of other parameters are as follows:

Nonlocal parameter: $\mu = 0, 1, 2\ \text{nm}^2$,

Transverse functionally gradient index: $n_z = 0, 1, 5$,

Axially functionally gradient index: $n_x = 0, 1, 2$,

Nonuniformity parameters: $0 \le \beta_h, \beta_b \le 0.5$,

Slenderness ratio: $5 \le L/h_0 \le 20$,

Temperatures: $T_0 = 0, \quad T_m - T_0 = 5(\text{K}), \quad 0(\text{K}) \le dT \le 100(\text{K})$, and $\nu = 0.3$.

2.4.1 Convergence Study and Validation

Regarding the convergence of GDQM, a convergence test has been performed to obtain a sufficient number of grid points required for the accuracy of up to four decimal places. The MATLAB codes developed for frequencies were run to compute the frequency parameter Ω, taking $n = 10 - 18$ for different sets of values of beam parameters. It has been noticed that the difference $\left|\Omega^{(n+1)} - \Omega^{(n)}\right| \le 0.00005$ for $n \ge 14$ for the S-S beam, while $n \ge 16$ for C-C, C-S and C-F beams in all three modes. The results have been shown in Table 2.2 for a specified beam with $\mu = 2,\ n_z = 5,\ n_x = 2,\ L/h_0 = 20,\ (\beta_h, \beta_b) = (0.5, 0.5)$

TABLE 2.1

Thermo-Mechanical Properties of Metal and Ceramic Constituent's Materials

Material	Properties	P_{-1}	P_0	P_1	P_2	P_3
SUS_3O_4	$E(\text{Pa})$	0	201.04e+9	3.079e-4	–6.534e-7	0
	$\alpha(\text{K}^{-1})$	0	12.33e-6	8.086e-4	0	0
	$\rho(\text{kg/m}^3)$	0	8166	0	0	0
	$k(\text{W/mK})$	0	15.379	–1.264e-3	2.092e-6	–7.223e-10
Si_3N_4	$E(\text{Pa})$	0	348.43e+9	–3.07e-4	2.16e-7	–8.946e-11
	$\alpha(\text{K}^{-1})$	0	5.8723e-6	9.095e-4	0	0
	$\rho(\text{kg/m}^3)$	0	2370	0	0	0
	$k(\text{W/mK})$	0	13.7230	–1.032e-3	5.466e-7	–7.876e-11

TABLE 2.2
Convergence of GDQM for Ω of BD-FG Nonuniform Nanobeams

	C-C			C-S			S-S		
n	Ω_1	Ω_2	Ω_3	Ω_1	Ω_2	Ω_3	Ω_1	Ω_2	Ω_3
10	5.9213	13.5954	21.5444	4.0064	11.9256	20.2954	2.2851	9.6951	17.8771
11	**5.9213**	13.5859	21.6287	**4.0064**	11.9247	20.2837	**2.2851**	9.6948	17.8811
12	—	13.5842	21.5228	—	11.9248	20.2648	—	9.695	17.872
13	—	13.5843	21.5097	—	11.9249	20.2631	—	**9.695**	17.8727
14	—	**13.5843**	21.5112	—	**11.9249**	20.2633	—	—	17.8727
15	—	—	21.5111	—	—	20.2632	—	—	17.8726
16	—	—	**21.5111**	—	—	**20.2632**	—	—	**17.8726**
17	—	—	—	—	—	—	—	—	—
18	—	—	—	—	—	—	—	—	—

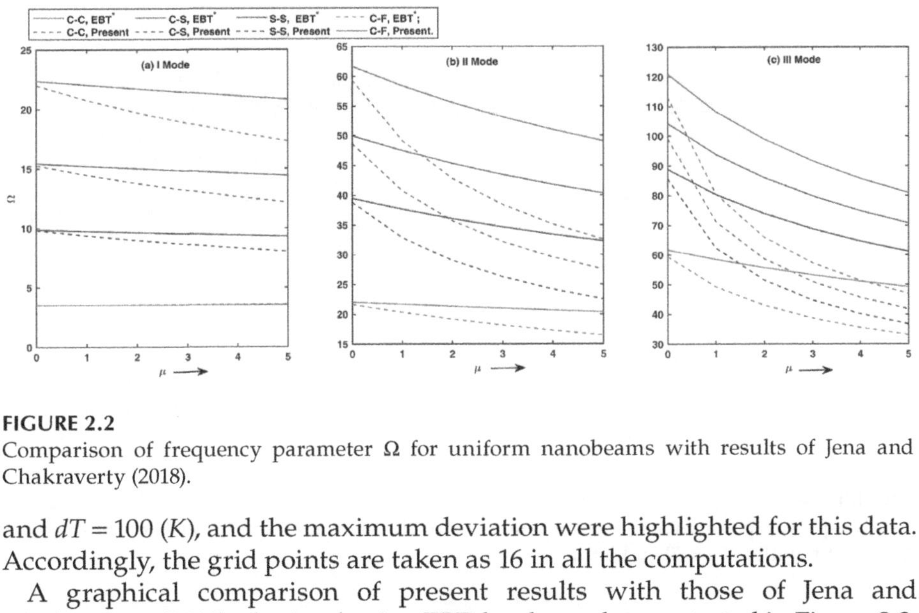

FIGURE 2.2
Comparison of frequency parameter Ω for uniform nanobeams with results of Jena and Chakraverty (2018).

and $dT = 100$ (K), and the maximum deviation were highlighted for this data. Accordingly, the grid points are taken as 16 in all the computations.

A graphical comparison of present results with those of Jena and Chakraverty (2018), obtained using EBT, has been demonstrated in Figure 2.2, for varying values of the nonlocal parameter in the absence of thermal field $(T = 0, \alpha = 0)$. A close agreement of the results shows the versatility of the present technique.

2.4.2 Vibration Analysis

Selective data on the frequency parameter of BD-FG nonuniform nanobeam subjected to NTD have been presented in Figures 2.3–2.11. It is clear that the values of Ω for C-C nanobeam are higher than the corresponding C-S,

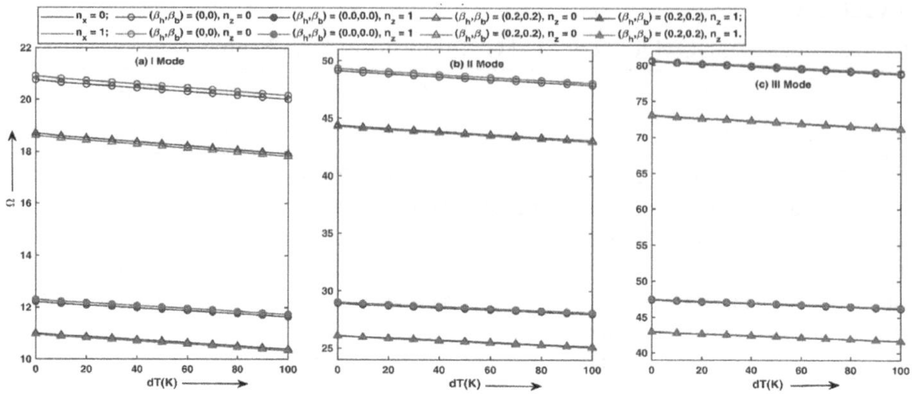

FIGURE 2.3
The frequency parameter Ω versus temperature difference dT for BD-FG nonuniform C-C nanobeam with $L/h_0 = 20$ and $\mu = 1$.

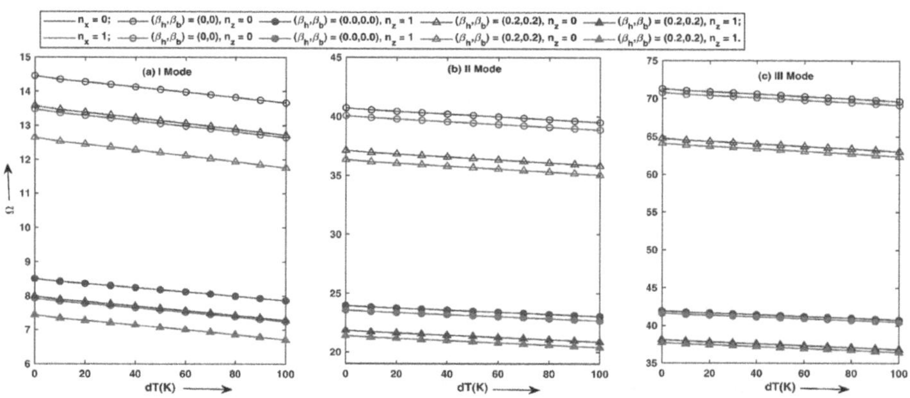

FIGURE 2.4
The frequency parameter Ω versus temperature difference dT for BD-FG nonuniform C-S nanobeam with $L/h_0 = 20$ and $\mu = 1$.

and S-S nanobeams. The frequency parameter Ω is found to decrease with the increasing value of the nonlocal parameter μ for all the boundary conditions. Moreover, the increasing value of TFG n_z has a decreasing effect on the frequency parameter. Figures 2.3–2.5 show the graphs of frequency parameter Ω versus temperature difference dT, taking $n_z = 0, 1$, $n_x = 0, 1$ and $(\beta_h, \beta_b) = (0.0, 0.0), (0.2, 0.2)$ with $\mu = 1$, $L/h_0 = 20$. It is observed that the increasing values of dT soften the beam for all the boundary conditions and modes. This effect is highly noticeable for a nonuniform FG nanobeam, i.e., for

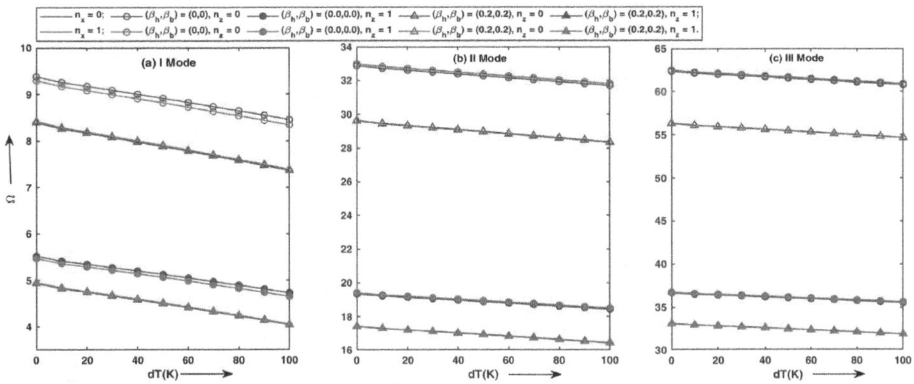

FIGURE 2.5
The frequency parameter Ω versus temperature difference dT for BD-FG nonuniform S-S nanobeam with $L/h_0 = 20$ and $\mu = 1$.

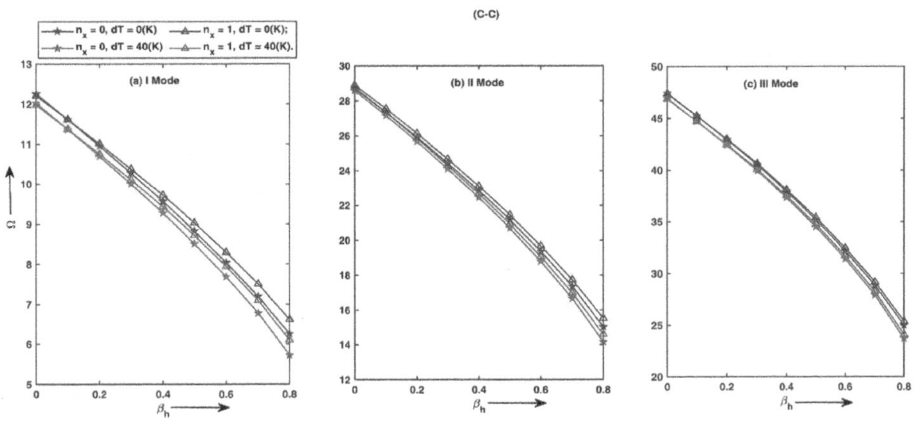

FIGURE 2.6
The frequency parameter Ω versus nonuniformity parameter along the thickness β_h for BD-FG C-C nanobeam with $L/h_0 = 20$ and $\mu = 1$, $n_z = 1$ and $\beta_b = 0.2$.

higher values of n_z and (β_h, β_b). Moreover, the consideration of bi-directional gradation by increasing the values of AFG n_x further intensifies this effect. This effect increases with the increasing modes and edge fixity condition.

Figures 2.6–2.8 depict the variation of the frequency parameter Ω with the nonuniformity parameter along the thickness β_h taking $dT = 0, 40(\text{K})$, $n_x = 0, 1$ with $\mu = 1$, $n_z = 1$, $L/h_0 = 20$, and $\beta_b = 0.2$. The value of the frequency parameter Ω decreases with the increasing value of β_h, for all the boundary

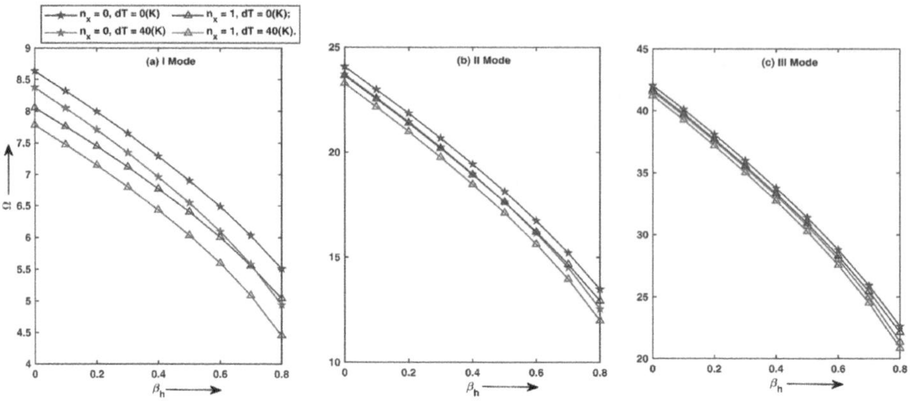

FIGURE 2.7
The frequency parameter Ω versus nonuniformity parameter along the thickness β_h for BD-FG C-S nanobeam with $L/h_0 = 20$ and $\mu = 1$, $n_z = 1$ and $\beta_b = 0.2$.

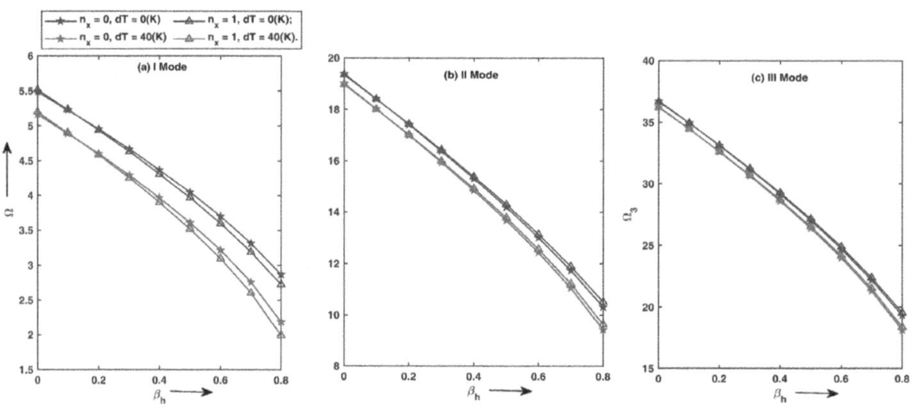

FIGURE 2.8
The frequency parameter Ω versus nonuniformity parameter along the thickness β_h for BD-FG S-S nanobeam with $L/h_0 = 20$ and $\mu = 1$, $n_z = 1$ and $\beta_b = 0.2$.

conditions and mode numbers. The rate of decay increases for higher values of dT. This effect further increases with the increasing values of n_x, except for the S-S beam in the second and third modes and the C-C beam in all modes. The effect of β_h on Ω increases with the edge fixity condition and the number of modes.

The variation in the frequency parameter Ω of the BD-FG nanobeam with varying values of AFG n_x has been presented in Figures 2.9–2.11, taking $\mu = 1,2$ and $(\beta_h, \beta_b) = (0,0),(0.2,0.2),(0.3,0.3),(0.5,0.5)$ with $L/h_0 = 20$ and

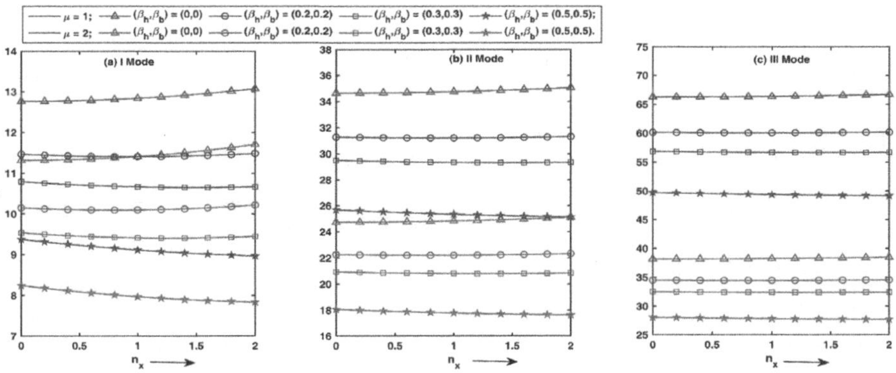

FIGURE 2.9
The frequency parameter Ω versus AFG n_x for BD-FG nonuniform C-C nanobeam with $L/h_0 = 20$, $dT = 40(\text{K})$, $n_z = 1$.

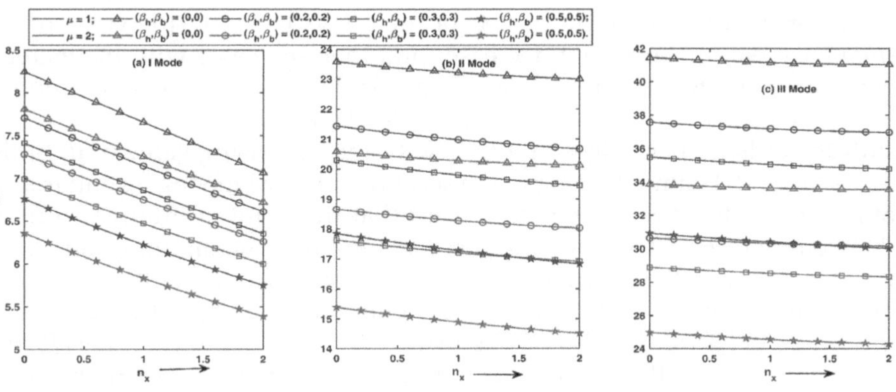

FIGURE 2.10
The frequency parameter Ω versus AFG n_x for BD-FG nonuniform C-S nanobeam with $L/h_0 = 20$, $dT = 40(\text{K})$, $n_z = 1$.

$n_z = 1$, $dT = 40(\text{K})$ for all boundary conditions and the first three modes. The graphs show that the value of the frequency parameter Ω decreases continuously with the increasing value of n_x, for C-S (Figure 2.10) nanobeam. This effect increases with the increase in the value of (β_h, β_b), while it decreases with the increase in the nonlocal parameter and modes.

In the case of C-C (Figure 2.9) and S-S (Figure 2.11) beams, the behaviour of Ω depends upon the values of (β_h, β_b) and modes. For a uniform beam, i.e., $(\beta_h, \beta_b) = (0, 0)$, the frequency parameter Ω increases monotonically with the increasing value of n_x, for both the boundary conditions and all three

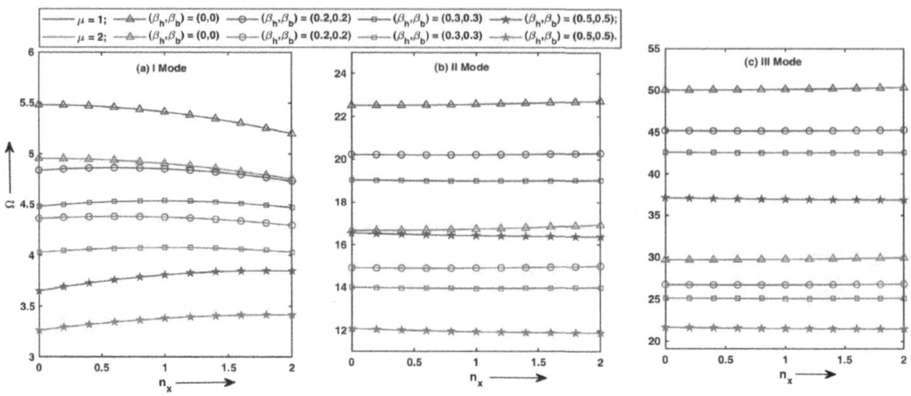

FIGURE 2.11
The frequency parameter Ω versus AFG n_x for BD-FG nonuniform S-S nanobeam with $L/h_0 = 20$, $dT = 40(\text{K})$, $n_z = 1$.

modes except for the S-S beam vibrating in the first mode, where it is just the reverse. For a nonuniform C-C beam $((\beta_h, \beta_b) > (0,0))$, the value of the frequency parameter Ω first decreases and then increases with the increasing value of n_x, for all three modes. In particular, there is a point of minima in the vicinity of $n_x = 1$ when $(\beta_h, \beta_b) = (0.2, 0.2)$, $\mu = 0$ and $dT = 40(\text{K})$, which shift towards higher values of $n_x (> 1)$ with the increasing value of μ. In the case of the S-S beam vibrating in the second and third modes, a similar kind of behaviour has been observed, except that the point of minima shifts in the vicinity of $n_x = 0.6$. However, there is a point of maximum in the vicinity of $n_x = 0.6$, in the first mode of the S-S beam. These points of maxima or minima seem to shift towards the higher values of n_x with the increasing values of $(\beta_h, \beta_b) < (0.4, 0.4)$. The further increase in the values of (β_h, β_b), gradually changes the nature of Ω with increasing value of n_x and becomes just the reverse for both the boundary conditions and all three modes when $(\beta_h, \beta_b) = (0.5, 0.5)$.

2.5 Summary

In summary, this chapter systematically outlines the development of a mathematical framework to analyse the vibration behaviour of BD-FG nonuniform nanobeams under thermal loading. Additionally, it establishes the foundational understanding of the nonlocal model used throughout the subsequent chapters of this book. The study demonstrates that vibration characteristics are highly sensitive to boundary conditions, temperature variations, geometric parameters, and material gradation. An increase in temperature generally leads to a reduction in natural frequency, with this effect being more pronounced in FG nonuniform beams compared to uniform ones. Material gradation and geometric nonuniformity play a significant role in determining the extent of frequency variation, depending on specific design configurations. The findings offer valuable insights for the design and optimization of nanoscale structural components in high-temperature applications.

References

Chakraverty, S., and Laxmi Behera. 2014. "Free Vibration of Rectangular Nanoplates Using Rayleigh-Ritz Method." *Physica E: Low-Dimensional Systems and Nanostructures* 56:357–63. doi: 10.1016/j.physe.2013.08.014.

Eringen, A. Cemal. 1983. "On Differential Equations of Nonlocal Elasticity and Solutions of Screw Dislocation and Surface Waves." *Journal of Applied Physics* 54(9):4703–10. doi: 10.1063/1.332803.

Eringen, A. Cemal, and D. G. B. Edelen. 1972. "On Nonlocal Elasticity." *International Journal of Engineering Science* 10(3):233–48. doi: 10.1016/0020-7225(72)90039-0.

Jena, Subrat Kumar, and S. Chakraverty. 2018. "Free Vibration Analysis of Euler–Bernoulli Nanobeam Using Differential Transform Method." *International Journal of Computational Materials Science and Engineering* 7(3). doi: 10.1142/S2047684118500203.

Khaniki, Hossein Bakhshi, and Sundaramoorthy Rajasekaran. 2018. "Mechanical Analysis of Non-Uniform Bi-Directional Functionally Graded Intelligent Micro-Beams Using Modified Couple Stress Theory." *Materials Research Express* 5(5). doi: 10.1088/2053-1591/aabe62.

Kumar, Rajeev, B. K. Mishra, and S. C. Jain. 2008. "Thermally Induced Vibration Control of Cylindrical Shell Using Piezoelectric Sensor and Actuator." *International Journal of Advanced Manufacturing Technology* 38(5–6):551–62. doi: 10.1007/s00170-007-1076-y.

Mahinzare, Mohammad, Mohammad Mostafa Barooti, and Majid Ghadiri. 2018. "Vibrational Investigation of the Spinning Bi-Dimensional Functionally Graded (2-FGM) Micro Plate Subjected to Thermal Load in Thermal Environment." *Microsystem Technologies* 24(3):1695–711. doi: 10.1007/s00542-017-3544-0.

Mathew, Jinu, Josny Joy, and Soney C. George. 2019. "Potential Applications of Nanotechnology in Transportation: A Review." *Journal of King Saud University - Science* 31(4):586–94. doi: 10.1016/j.jksus.2018.03.015.

Mirjavadi, Seyed Sajad, Behzad Mohasel Afshari, Navvab Shafiei, A. M. S. Hamouda, and Mohammad Kazemi. 2017. "Thermal Vibration of Two-Dimensional Functionally Graded (2D-FG) Porous Timoshenko Nanobeams." *Steel and Composite Structures* 25(4):415–26.

Mirjavadi, Seyed Sajad, Abdolhamid Matin, Navvab Shafiei, Samira Rabby, and Behzad Mohasel Afshari. 2017. "Thermal Buckling Behaviour of Two-Dimensional Imperfect Functionally Graded Microscale-Tapered Porous Beam." *Journal of Thermal Stresses* 40(10):1201–14. doi: 10.1080/01495739.2017.1332962.

Nejad, Mohammad Zamani, and Amin Hadi. 2016. "Eringen's Non-Local Elasticity Theory for Bending Analysis of Bi-Directional Functionally Graded Euler-Bernoulli Nano-Beams." *International Journal of Engineering Science* 106:1–9. doi: 10.1016/j.ijengsci.2016.05.005.

Nemat-Alla, Mahmoud. 2003. "Reduction of Thermal Stresses by Developing Two-Dimensional Functionally Graded Materials." *International Journal of Solids and Structures* 40:7339–56. doi: 10.1016/j.ijsolstr.2003.08.017.

Norouzzadeh, A., and R. Ansari. 2017. "Finite Element Analysis of Nano-Scale Timoshenko Beams Using the Integral Model of Nonlocal Elasticity." *Physica E: Low-Dimensional Systems and Nanostructures* 88(October 2016):194–200. doi: 10.1016/j.physe.2017.01.006.

Rajasekaran, Sundaramoorthy, and Hossein Bakhshi Khaniki. 2018. "Bending, Buckling and Vibration Analysis of Functionally Graded Non-uniform Nanobeams via Finite Element Method." *Journal of the Brazilian Society of Mechanical Sciences and Engineering* 40(11). doi: 10.1007/s40430-018-1460-6.

Reddy, J. N. 2008. "Theory and Analysis of Elastic Plates and Shells." *CRC Press* 88(10):827–827. doi: 10.1002/zamm.200890020.

Sahmani, Saeid, and Babak Safaei. 2019. "Nonlinear Free Vibrations of Bi-Directional Functionally Graded Micro/Nano-Beams Including Nonlocal Stress and Microstructural Strain Gradient Size Effects." *Thin-Walled Structures* 140(November 2018):342–56. doi: 10.1016/j.tws.2019.03.045.

Shafiei, Navvab, and Mohammad Kazemi. 2017. "Buckling Analysis on the Bi-Dimensional Functionally Graded Porous Tapered Nano-/Micro-Scale Beams." *Aerospace Science and Technology* 66:1–11. doi: 10.1016/j.ast.2017.02.019.

Shafiei, Navvab, Seyed Sajad Mirjavadi, Behzad MohaselAfshari, Samira Rabby, and Mohammad Kazemi. 2017. "Vibration of Two-Dimensional Imperfect Functionally Graded (2D-FG) Porous Nano-/Micro-Beams." *Computer Methods in Applied Mechanics and Engineering* 322:615–32. doi: 10.1016/j.cma.2017.05.007.

Shahrjerdi, A., F. Mustapha, M. Bayat, and D. L. A. Majid. 2011. "Free Vibration Analysis of Solar Functionally Graded Plates with Temperature-Dependent Material Properties Using Second Order Shear Deformation Theory." *Journal of Mechanical Science and Technology*. doi: 10.1007/s12206-011-0610-x.

Shu, Chang. 2011. *Differential Quadrature and Its Application in Engineering*. Springer London.

Sun, Rujie, Qinyu Li, Jianfei Yao, Fabrizio Scarpa, and Jonathan Rossiter. 2020. "Tunable, Multi-Modal, and Multi-Directional Vibration Energy Harvester Based on Three-Dimensional Architected Metastructures." *Applied Energy* 264(January):114615. doi: 10.1016/j.apenergy.2020.114615.

Tang, Ye, Xiaofei Lv, and Tianzhi Yang. 2019. "Bi-Directional Functionally Graded Beams: Asymmetric Modes and Nonlinear Free Vibration." *Composites Part B: Engineering* 156:319–31. doi: 10.1016/j.compositesb.2018.08.140.

Trinh, Luan C., Thuc P. Vo, Huu-Tai Thai, and Trung-Kien Nguyen. 2018. "Size-Dependent Vibration of Bi-Directional Functionally Graded Microbeams with Arbitrary Boundary Conditions." *Composites Part B* 134:225–45. doi: 10.1016/j.compositesb.2017.09.054.

3

Influence of In-Plane Load on the Thermal Buckling and Vibration Characteristics of BD-FG Nonuniform Nanobeams

3.1 Overview

In various micro/nano-devices, the nanostructures embedded in a thermal environment may be subjected to different types of mechanical loads, such as in-plane transverse load, point load, electrical load, rotating force, etc. (Palaniselvam and Baek 2015; Samorì et al. 2015). Nowadays, the utilization of advanced nanostructures will help in enhancing the sensitivity of these devices (Shen et al. 2012; Nemat-Alla 2003). Further, the consideration of cross-sectional nonuniformity in nanostructures increases the efficiency of these systems. Owing to the merits of non-homogeneity in geometry and material, considerable attention has been given to determine the response of FG nonuniform nanobeams and nanoplates under the influence of thermal load, together with in-plane load. In this context, some significant contributions on one-directional FG nanostructures under the combined effect of thermal and in-plane loads have been reported in the literature as (Ebrahimi, Barati, and Haghi 2017; Ebrahimi and Barati 2017; Rajasekaran and Bakhshi 2017; Shafiei, Ghadiri, and Mahinzare 2017). In the case of BD-FG nanostructures, an analysis for the buckling behaviour of the BD-FG tapered Euler–Bernoulli nanobeam has been presented by Shafiei and Kazemi (2017) on the basis of MCST, where the beam is tapered linearly in both height and width directions. GDQM has been used to solve the governing equations for the clamped boundary condition. Khaniki and Rajasekaran (2018) investigated the mechanical behaviour of a BD-FG porous Euler-Bernoulli nanobeam of a general varying cross-section using MCST. Mixed FEM is used as a solution technique. The effect of a rotating radial in-plane load on the natural frequencies of a BD-FG micro circular plate exposed to thermal load resting on Winkler–Pasternak foundation has been analysed by Mahinzare, Barooti, and Ghadiri (2018) on the basis of FSDT and MCST. The material properties of the microplate are supposed to change across the thickness and radius directions of the plate. DQM is used to compute the frequencies of the considered plate.

DOI: 10.1201/9781003583806-3

In view of the above, the combined effect of mechanical and thermal in-plane loads on the vibration behaviour of the BD-FG nanobeam having a nonuniform cross-section along the length due to linear variation in thickness has been investigated in this chapter. The beam is exposed to a uniform mechanical load and a nonlinearly varying temperature field across the thickness. The material properties are supposed to be temperature-dependent and vary in the thickness as well as the length direction according to a power-law model and an exponential function, respectively. The governing differential equations of motion for the present problem have been obtained using Hamilton's principle on the basis of Eringen's nonlocal theory and FSDT. The GDQM has been employed to obtain the numerical solution for frequency parameters. A parametric study has been performed to examine the combined effect of thermal and in-plane load on the frequency parameter of BD-FG tapered nanobeams, for the first three modes. By allowing the frequency parameter to approach zero, the critical buckling loads in compression have been computed.

3.2 Mathematical Formulation

Referred to a Cartesian coordinate system (x, y, z), consider a BD-FG nonuniform nanobeam of length L, width b, and variable thickness $h(x)$, subjected to uniform in-plane load. N_0 in a nonlinear thermal environment. The middle plane is $z = 0$ with the origin at the end $x = 0$. The top $(z = +h/2)$ and bottom $(z = -h/2)$ surfaces of the beam are fully ceramic and metallic, respectively. The nonuniformity in the cross-section of the beam arises due to thickness variation along the length as $h(x) = h_0\left(1 - \beta_h \, {x}/{L}\right)$, where β_h denotes the rate of the cross-sectional change along the thickness and h_0 is the thickness of the beam at the origin, as shown in Figure 3.1.

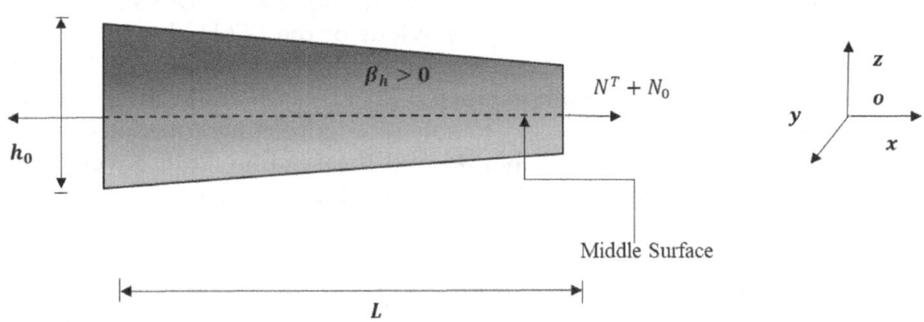

FIGURE 3.1
Cross-sectional view of the BD-FG nonuniform nanobeam subjected to the in-plane load in the Cartesian coordinate system.

3.2.1 Energy Relations

In addition to Chapter 2, the energy stored in the nanobeam due to the applied uniform mechanical in-plane load N_0 and thermal in-plane load N^T arises due to nonlinear temperature variation across the thickness is given by,

$$U_{axial} = \frac{1}{2}\int_{-b/2}^{b/2}\left(\int_0^L \left(N^T + N_0\right)\frac{\partial^2 w_0}{\partial x^2}dx\right)dy \tag{3.2.1}$$

where thermal load against temperature variation dT is

$$N^T = \int_{-h(x)/2}^{h(x)/2} E(x,z,T)\alpha(x,z,T)\left(T(z)-T_0\right)dz,\ T_0 \text{ is the initial temperature.}$$

Taking the first variation of equation (3.2.1) and integrating with respect to y, gives

$$\delta U_{axial} = \int_0^L b\left(N^T + N_0\right)\frac{\partial w_0}{\partial x}\frac{\partial(\delta w_0)}{\partial x}dx \tag{3.2.2}$$

3.2.2 Nonlocal Governing Equations of Motion

Substituting the expressions for potential energy (2.2.10), the kinetic energy (2.2.12) and work done by combined thermal and in-plane loads (3.2.2) into Hamilton's energy principle for $\delta L = \delta K - \delta U - \delta U_{axial}$, the Euler-Lagrange equations are obtained as:

$$\frac{\partial N_{xx}^{nl}}{\partial x} = I_0\frac{\partial^2 u_0}{\partial t^2} + I_1\frac{\partial^2 \psi_x}{\partial t^2} \tag{3.2.3}$$

$$\frac{\partial M_{xx}^{nl}}{\partial x} - Q_{xz}^{nl} = I_1\frac{\partial^2 u_0}{\partial t^2} + I_2\frac{\partial^2 \psi_x}{\partial t^2} \tag{3.2.4}$$

$$\frac{\partial Q_{xz}^{nl}}{\partial x} + \left(N^T + N_0\right)\frac{\partial^2 w_0}{\partial x^2} = I_0\frac{\partial^2 w_0}{\partial t^2} \tag{3.2.5}$$

Now, using the expression for nonlocal stresses and moments (2.2.9), the nonlocal governing equations of motion (3.2.3)–(3.2.5) for the considered beam are obtained as follows:

$$A_{xx}\frac{\partial^2 u_0}{\partial x^2} + B_{xx}\frac{\partial^2 \psi_x}{\partial x^2} + \frac{\partial A_{xx}}{\partial x}\frac{\partial u_0}{\partial x} + \frac{\partial B_{xx}}{\partial x}\frac{\partial \psi_x}{\partial x} = \left(1-\mu\frac{\partial^2}{\partial x^2}\right)\left(I_0\frac{\partial^2 u_0}{\partial t^2} + I_1\frac{\partial^2 \psi_x}{\partial t^2}\right) \tag{3.2.6}$$

$$B_{xx}\frac{\partial^2 u_0}{\partial x^2}+D_{xx}\frac{\partial^2 \psi_x}{\partial x^2}+\frac{\partial B_{xx}}{\partial x}\frac{\partial u_0}{\partial x}+\frac{\partial D_{xx}}{\partial x}\frac{\partial \psi_x}{\partial x}-C_{xz}\left(\frac{\partial w_0}{\partial x}+\psi_x\right)$$
$$=\left(1-\mu\frac{\partial^2}{\partial x^2}\right)\left(I_1\frac{\partial^2 u_0}{\partial t^2}+I_2\frac{\partial^2 \psi_x}{\partial t^2}\right) \tag{3.2.7}$$

$$C_{xz}\frac{\partial^2 w_0}{\partial x^2}+C_{xz}\frac{\partial \psi_x}{\partial x}+\frac{\partial C_{xz}}{\partial x}\left(\frac{\partial w_0}{\partial x}+\psi_x\right)+\left(1-\mu\frac{\partial^2}{\partial x^2}\right)\frac{\partial}{\partial x}\left(\left(N^T+N_0\right)\frac{\partial w_0}{\partial x}\right)$$
$$=\left(1-\mu\frac{\partial^2}{\partial x^2}\right)\left(I_0(x)\frac{\partial^2 w_0}{\partial t^2}\right) \tag{3.2.8}$$

For harmonic vibration, assume that $u_0(x,t)=\tilde{u}_0 e^{i\omega t}$, $w_0(x,t)=\tilde{w}_0 e^{i\omega t}$, $\psi_x(x,t)=\tilde{\psi}_x e^{i\omega t}$ and using non-dimensional variables $\bar{w}_0=\frac{\tilde{w}_0}{L}$, $\bar{u}_0=\frac{\tilde{u}_0}{L}$, $\bar{\psi}_x=\tilde{\psi}_x$, $X=\frac{x}{L}$, $H=\frac{h_0}{L}$ and $Z=\frac{z}{h}$, the nondimensionalized form of equations (3.2.6)–(3.2.8) is

$$U_0\frac{d^2\bar{u}_0}{dX^2}+U_1\frac{d^2\bar{\psi}_x}{dX^2}+\frac{dU_0}{dX}\frac{d\bar{u}_0}{dX}+\frac{dU_1}{dX}\frac{d\bar{\psi}_x}{dX}=-\hat{\Omega}^2\left(1-\gamma^2\frac{d^2}{dX^2}\right)\left[V_0\bar{u}_0+V_1\bar{\psi}_x\right] \tag{3.2.9}$$

$$U_1\frac{d^2\bar{u}_0}{dX^2}+U_2\frac{d^2\bar{\psi}_x}{dX^2}+\frac{dU_1}{dX}\frac{d\bar{u}_0}{dX}+\frac{dU_2}{dX}\frac{d\bar{\psi}_x}{dX}-U_3\left(\frac{d\bar{w}_0}{dX}+\bar{\psi}_x\right)$$
$$=-\hat{\Omega}^2\left(1-\gamma^2\frac{d^2}{dX^2}\right)\left[V_1\bar{u}_0+V_2\bar{\psi}_x\right] \tag{3.2.10}$$

$$U_3\frac{d^2\bar{w}_0}{dX^2}+U_3\frac{d\bar{\psi}_x}{dX}+\frac{dU_3}{dX}\left(\frac{d\bar{w}_0}{dX}+\bar{\psi}_x\right)+\left(1-\gamma^2\frac{d^2}{dX^2}\right)\left(\frac{d}{dX}\left(P_1\frac{d\bar{w}_0}{dX}\right)\right)$$
$$+\frac{H^3}{12}\left(1-\gamma^2\frac{\partial^2}{\partial X^2}\right)\left(N^*\frac{d^2\bar{w}_0}{dX^2}\right)=-\hat{\Omega}^2\left(1-\gamma^2\frac{d^2}{dX^2}\right)\left[V_0\bar{w}_0\right] \tag{3.2.11}$$

Where,

$$U_0=e^{n_x X}\left(1-\beta_h X\right)\bar{A}_{xx}H,\quad U_1=e^{n_x X}\left(1-\beta_h X\right)^2\bar{B}_{xx}H^2,$$

$$U_2 = e^{n_x X}(1-\beta_h X)^3 \bar{D}_{xx} H^3, \quad U_3 = e^{n_x X}(1-\beta_h X)\bar{C}_{xz} H,$$

$$V_0 = e^{n_x X}(1-\beta_h X)\bar{I}_0 H, \quad V_1 = e^{n_x X}(1-\beta_h X)^2 \bar{I}_1 H^2,$$

$$V_2 = e^{n_x X}(1-\beta_h X)^3 \bar{I}_2 H^3; \quad P_1 = e^{2n_x X}(1-\beta_h X) H\alpha_0^c \bar{N}^T,$$

$$\bar{N}^T = \frac{N^T}{bhE_0^c}, N^* = \frac{N_0 L^2}{\bar{D}_{xx}}, \hat{\Omega}^2 = \frac{\omega^2 L^2 \rho_0^c}{E_0^c}, \left(\bar{A}_{xx}, \bar{B}_{xx}, \bar{D}_{xx}\right) = \left(\frac{A_{xx}}{bhE_0^c}, \frac{B_{xx}}{bh^2 E_0^c}, \frac{D_{xx}}{bh^3 E_0^c}\right),$$

$$\bar{C}_{xz} = \frac{C_{xz}}{bhE_0^c}, \text{ and } \left(\bar{I}_0, \bar{I}_1, \bar{I}_2\right) = \left(\frac{I_0}{bh\rho_0^c}, \frac{I_1}{bh^2\rho_0^c}, \frac{I_2}{bh^3\rho_0^c}\right).$$

For harmonic vibration and introducing the non-dimensional variables, the nondimensionalized form of equations (3.2.9)–(3.2.11) is obtained and solved approximately using GDQM.

3.3 Solution Procedure

Now, employing GDQM, the discretization of governing equations (3.2.9)–(3.2.11) is performed at internal grid points. $X_j\left(j = 2, 3, \ldots, (n-1)\right)$ together with the boundary condition for C-C beam gives a complete set of $3n$ equations in $3n$ and unknowns $\bar{u}_0(X_j)$, $\bar{w}_0(X_j)$ and $\bar{\psi}_x(x_j)$, $(j = 1, 2, \ldots, n)$ and can be written in matrix form as

$$\begin{bmatrix} [\mathbf{K_{dd}}] & [\mathbf{K_{db}}] \\ [\mathbf{K_{bd}^{CC}}] & [\mathbf{K_{bb}^{CC}}] \end{bmatrix} \begin{Bmatrix} \{\mathbf{F_d}\} \\ \{\mathbf{F_b}\} \end{Bmatrix} - \hat{\Omega}^2 \begin{bmatrix} [\mathbf{M_{dd}}] & [\mathbf{M_{db}}] \\ [\mathbf{M_{bd}^{CC}}] & [\mathbf{M_{bb}^{CC}}] \end{bmatrix} \begin{Bmatrix} \{\mathbf{F_d}\} \\ \{\boldsymbol{F_b}\} \end{Bmatrix} = [0] \quad (3.3.1)$$

Similarly, for C-S, and S-S boundary conditions, the governing equations are

$$\begin{bmatrix} [\mathbf{K_{dd}}] & [\mathbf{K_{db}}] \\ [\mathbf{K_{bd}^{CS}}] & [\mathbf{K_{bb}^{CS}}] \end{bmatrix} \begin{Bmatrix} \{\mathbf{F_d}\} \\ \{\mathbf{F_b}\} \end{Bmatrix} - \hat{\Omega}^2 \begin{bmatrix} [\mathbf{M_{dd}}] & [\mathbf{M_{db}}] \\ [\mathbf{M_{bd}^{CS}}] & [\mathbf{M_{bb}^{CS}}] \end{bmatrix} \begin{Bmatrix} \{\mathbf{F_d}\} \\ \{\mathbf{F_b}\} \end{Bmatrix} = [0] \quad (3.3.2)$$

$$\begin{bmatrix} [\mathbf{K_{dd}}] & [\mathbf{K_{db}}] \\ [\mathbf{K_{bd}^{SS}}] & [\mathbf{K_{bb}^{SS}}] \end{bmatrix} \begin{Bmatrix} \{\mathbf{F_d}\} \\ \{\mathbf{F_b}\} \end{Bmatrix} - \hat{\Omega}^2 \begin{bmatrix} [\mathbf{M_{dd}}] & [\mathbf{M_{db}}] \\ [\mathbf{M_{bd}^{SS}}] & [\mathbf{M_{bb}^{SS}}] \end{bmatrix} \begin{Bmatrix} \{\mathbf{F_d}\} \\ \{\mathbf{F_b}\} \end{Bmatrix} = 0 \quad (3.3.3)$$

3.4 Results and Discussion

In order to study the combined effect of mechanical and thermal in-plane loads on the vibration behaviour of the BD-FG nonuniform nanobeam, in-house MATLAB codes have been developed for computing the lowest three eigenvalues of equations (3.3.1)–(3.3.3) and are retained as the frequency parameter $\Omega = \left(\frac{\sqrt{12}L}{h_0} \right) \hat{\Omega}$, for the first three modes. In addition to the beam's parameter in Chapter 2, the in-plane load parameter is taken as: $-3 \leq N^* \leq 5$.

For the analysis, the numerical results for the frequency parameter Ω have been given in Tables 3.1 and 3.2 and Figures 3.2–3.10. It is clear from Table 3.1 that the value of the frequency parameter Ω for the nanobeam subjected to tensile the in-plane load is higher than the corresponding ones subjected to compressive the in-plane load. In reference to the boundary conditions, the

TABLE 3.1

The Values of Ω for the BD-FG Nonuniform Nanobeam with $\beta_h = 0.4$ and $L/h_0 = 20$

				$n_z = 0$			$n_z = 5$		
μ	N^*	n_x	Ω_1	S-S	C-S	C-C	S-S	C-S	C-C
0	−1	0	1	6.5614	12.0375	16.8228	2.8064	5.5117	7.8559
			2	29.7201	38.4548	46.5164	13.8702	18.0522	21.9213
			3	67.2658	79.0757	90.2217	31.7280	37.3449	42.6558
		2	1	6.5371	10.1765	16.6183	2.3644	4.5795	7.7637
			2	29.7026	36.7427	46.2112	13.8593	17.2273	21.7768
			3	67.2465	77.3747	89.8816	31.7175	36.5310	42.2102
	3	0	1	9.6706	14.5175	18.6081	4.9476	7.2113	9.0834
			2	33.0251	41.4432	48.9946	16.1436	20.1098	23.6301
			3	70.6542	82.2409	92.9825	34.0663	39.5299	44.5632
		2	1	9.6589	12.9814	18.2941	4.9383	6.5055	8.9186
			2	33.0108	39.8250	48.6521	16.1351	19.3489	23.4600
			3	70.6359	80.5887	92.6235	34.0564	38.7494	44.3884
2	−1	0	1	5.7549	10.4415	14.5651	2.7814	4.6704	6.6637
			2	21.4447	27.0318	32.3724	9.8141	12.4472	14.9706
			3	38.8807	44.6797	50.5664	17.9925	20.7085	23.4690
		2	1	5.7460	8.7634	14.4823	2.3546	3.8286	6.6395
			2	21.4512	25.8794	32.2533	9.8056	11.9000	14.9279
			3	38.8931	43.7730	50.4608	18.0042	20.2825	23.4306
	3	0	1	9.1666	13.5957	17.5682	4.7257	6.8341	8.7233
			2	25.9392	31.773	37.3141	12.8896	15.7003	18.3630
			3	44.7104	50.9456	57.1952	21.9968	25.0125	28.0240
		2	1	9.1980	12.3029	17.3295	4.7509	6.2635	8.6021
			2	25.9172	30.6715	37.0157	12.8815	15.1906	18.2044
			3	44.6719	50.0332	56.9383	21.9759	24.5847	27.8875

TABLE 3.2

The Values of Critical Buckling Load for BD-FG Nonuniform Nanobeams with $n_z = 5$, $L/h_0 = 20$ and $\beta_b = 0.2$, for All the Boundary Conditions

dT (K)			0			40		
β_k	μ	n_x	S-S	C-S	C-C	S-S	C-S	C-C
0.0	0	0	6.7258	13.6443	26.3691	5.9523	12.8645	25.5783
		2	6.7258	10.8526	36.4301	5.7436	9.8909	34.5852
	2	0	5.6170	9.7233	14.7349	4.8446	8.9471	13.9545
		2	6.1304	8.6443	17.2962	4.2056	7.6670	16.5766
0.4	0	0	3.2006	6.4835	12.5661	2.6087	5.9352	11.9705
		2	3.2013	5.0381	14.9825	2.5333	4.4133	14.0672
	2	0	2.6181	3.6115	6.9155	2.0299	3.4721	5.4721
		2	2.6596	3.1157	7.3647	2.0102	3.0188	6.8747

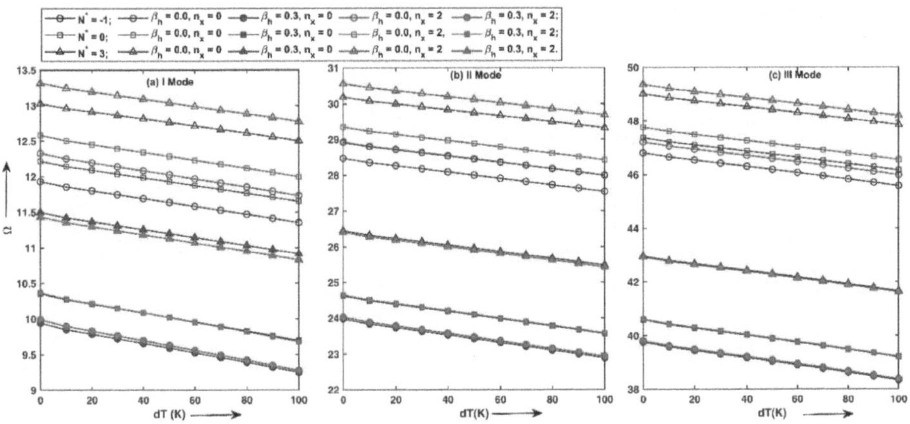

FIGURE 3.2
The frequency parameter Ω versus temperature difference dT for BD-FG nonuniform C-C nanobeam with $L/h_0 = 20$, $\mu = 1$ and $n_z = 1$.

values of the frequency parameter are in the order C-C>C-S>S-S beams. However, the effect of the in-plane load is more pronounced for the S-S beam as compared to C-S and C-C beams. The value of the frequency parameter Ω decreases as the value of the nonlocal parameter μ and TFG n_z. The effect of the in-plane load is more pronounced for higher values of μ, n_z. Although the influence of AFG n_x is complex and varies with the other parameters.

The variation in the frequency parameter Ω with temperature difference dT has been presented in Figures 3.2–3.4 for all three boundary conditions and modes, considering values of $n_x = 0, 2$, $\beta_h = 0.0, 0.3$ and $N^* = -1, 0, 3$. The frequency parameter Ω decreases with the increasing value of dT. This effect decreases as the in-plane load changes from compressive to tensile. The increasing values of the nonuniformity parameter β_h and AFG n_x enhance

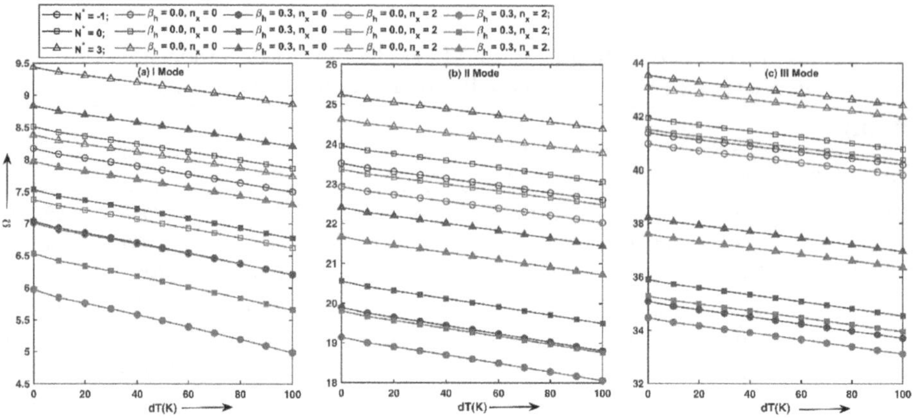

FIGURE 3.3
The frequency parameter Ω versus temperature difference dT for BD-FG nonuniform C-S nanobeam with $L/h_0 = 20$, $\mu = 1$ and $n_z = 1$.

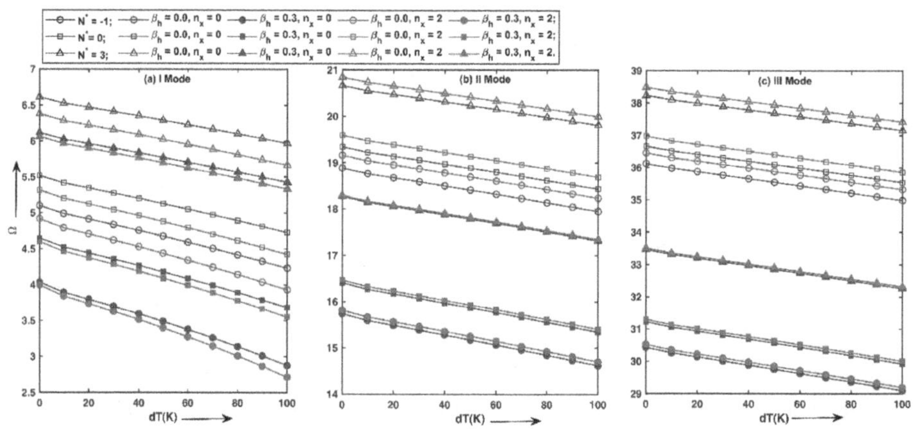

FIGURE 3.4
The frequency parameter Ω versus temperature difference dT for BD-FG nonuniform S-S nanobeam with $L/h_0 = 20$, $\mu = 1$ and $n_z = 1$.

the effect of dT on Ω. The rate of decrease in Ω with dT is highest for the S-S beam compared to the C-S and C-C beams. This effect further decreases with the increase in the number of modes.

Figures 3.5–3.7 show the variation of frequency parameter Ω for the BD-FG nanobeam with nonuniformity parameter along the thickness. β_h taking $dT = 0(\text{K}), 60(\text{K})$, $n_x = 0, 2$ and $N^* = -1, 0, 3$ for all three boundary conditions and modes. It has been observed that the value of Ω decreases with the

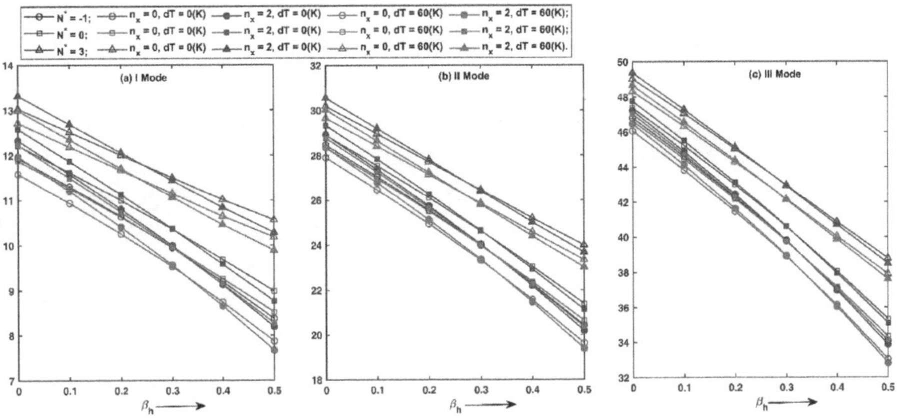

FIGURE 3.5
The frequency parameter Ω versus nonuniformity parameter along the thickness β_h for BD-FG nonuniform C-C nanobeam with $L/h_0 = 20$, $\mu = 1$ and $n_z = 1$.

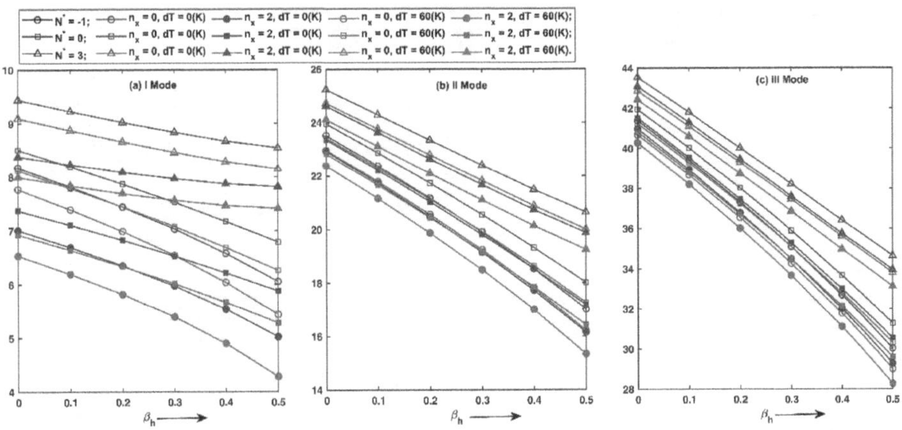

FIGURE 3.6
The frequency parameter Ω versus nonuniformity parameter along the thickness β_h for BD-FG nonuniform C-S nanobeam with $L/h_0 = 20$, $\mu = 1$ and $n_z = 1$.

increasing value of β_h, for all three modes and boundary conditions. The rate of decay is higher in the presence of compressive load as compared to tensile load. This effect further increases with the increasing values of n_x, except for the C-S and S-S vibrating in the mode beams. A high thermal environment increases the effect of β_h on Ω and highly noticeable on C-C beam as compared to S-S and C-S beams. The effect of β_h on Ω increases with the increase in the number of modes.

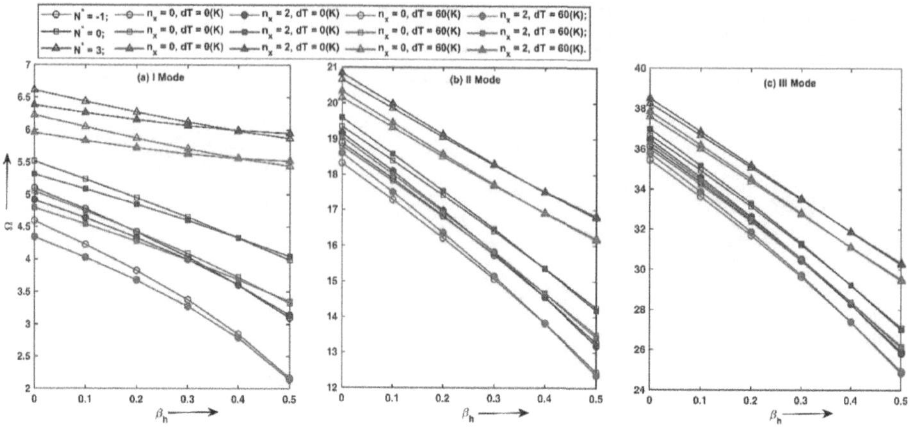

FIGURE 3.7
The frequency parameter Ω versus nonuniformity parameter along the thickness β_h for BD-FG nonuniform S-S nanobeam with $L/h_0 = 20$, $\mu = 1$ and $n_z = 1$.

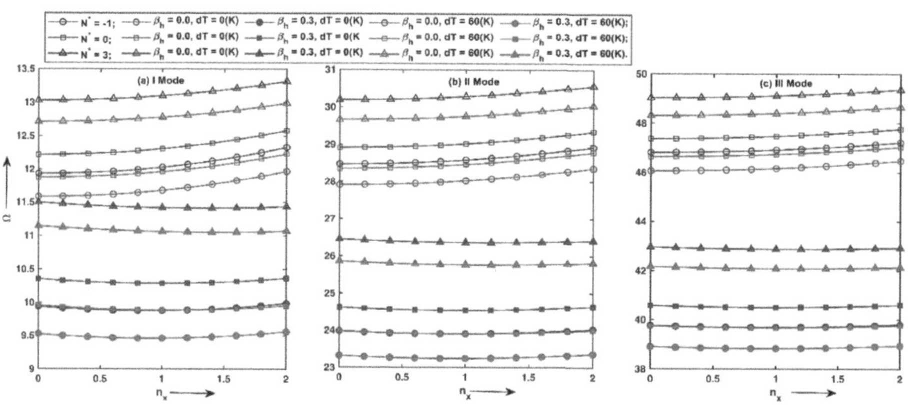

FIGURE 3.8
The frequency parameter Ω versus AFG n_x for BD-FG nonuniform C-C nanobeam with $L/h_0 = 20$, $\mu = 1$ and $n_z = 1$.

In Figures 3.8–3.10, frequency parameter Ω versus AFG n_x has been plotted, taking $\beta_h = 0.0, 0.3$, $N^* = -1, 0, 3$ and $dT = 0(\text{K}), 60(\text{K})$. For the C-S beam, the frequency parameter Ω decreases monotonically with increasing values of AFG n_x in all the modes, keeping other parameters fixed. The rate of decrease in Ω with n_x is higher for a uniform nanobeam, which slows down as the beam becomes thinner and thinner towards the outer edge. The effect of n_x on Ω is more noticeable for compressive load and a higher value of temperature difference dT.

In the case of uniform $(\beta_h = 0)$ C-C and S-S nanobeams, the frequency parameter Ω increases continuously with increasing value of n_x in all the

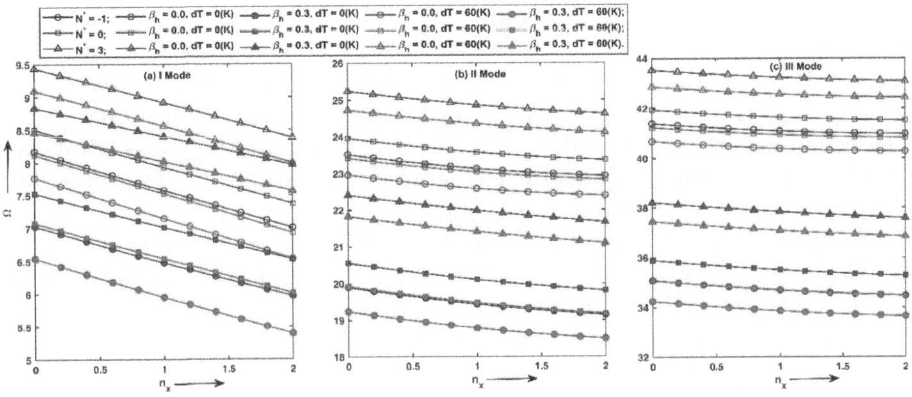

FIGURE 3.9
The frequency parameter Ω versus AFG n_x for BD-FG nonuniform C-S nanobeam with $L/h_0 = 20$, $\mu = 1$ and $n_z = 1$.

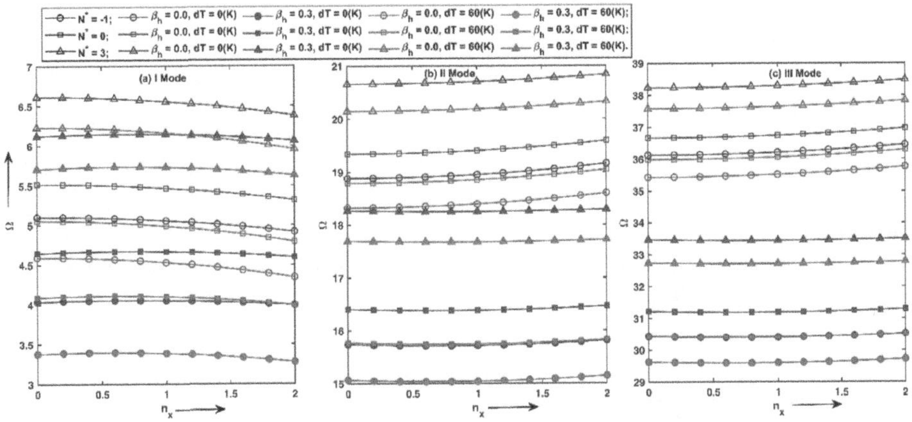

FIGURE 3.10
The frequency parameter Ω versus AFG n_x for BD-FG nonuniform C-C nanobeam with $L/h_0 = 20$, $\mu = 1$ and $n_z = 1$.

modes except for the S-S beam vibrating in the first mode, where it decreases. However, for the nonuniform nanobeams ($\beta_h > 0$), the frequency parameter Ω decreases first and then increases with the increasing value of n_x in all the modes. Hence, there exists a point of minima in the vicinity of $n_x = 1$ for both the beams, except for the S-S beam vibrating in the first mode of vibration, which attains maxima in the vicinity of $n_x = 1$. This point of minima or maxima shifts towards higher values of n_x as the in-plane load parameter varies from $N^* = -1$ to $N^* = 3$. In particular, for $N^* = -1$ the points of minima and maxima are $n_x = 0.8$ and $n_x = 0.6$ for C-C and S-S beams, respectively, while for $N^* = 3$ are $n_x = 1.4$ and $n_x = 0.8$ for C-C and S-S beams, respectively.

It is also worth noting that the change in the in-plane load from compressive to tensile, i.e., from $N^* = -1$ to $N^* = 3$, reverses the behaviour of Ω with n_x for the C-C beam in all the modes, and therefore the value of Ω at $n_x = 0$ is higher than that for the $n_x = 2$. These effects are more pronounced for higher values of dT. However, the effect of n_x on Ω is higher for C-S beam as compared to S-S and C-C beams. This effect is more noticeable on the first mode and decreases as the mode number increases.

In Figures 3.11–3.13, the variation of Ω with the in-plane load parameter N^* has been plotted, taking $n_x = 0, 2$, $\beta_h = 0.0, 0.3$, $dT = 0(\text{K}), 60(\text{K})$ for all

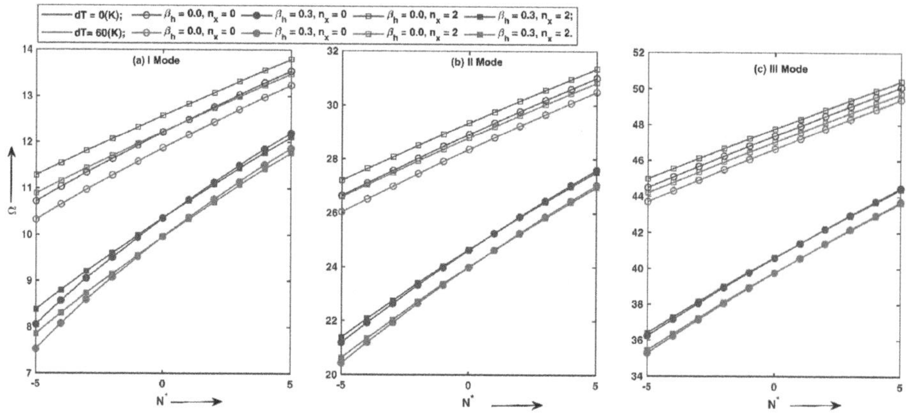

FIGURE 3.11
The frequency parameter Ω versus in-plane load parameter N^* for BD-FG nonuniform C-C nanobeam with $L/h_0 = 20$, $\mu = 1$ and $n_z = 1$.

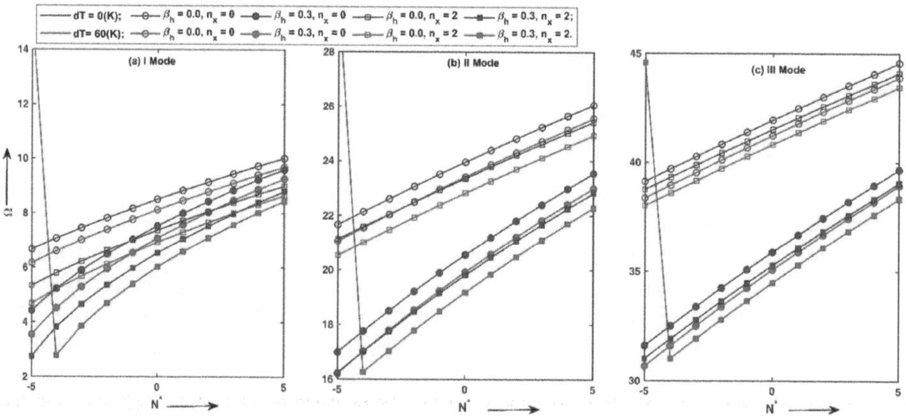

FIGURE 3.12
The frequency parameter Ω versus in-plane load parameter N^* for BD-FG nonuniform C-S nanobeam with $L/h_0 = 20$, $\mu = 1$ and $n_z = 1$.

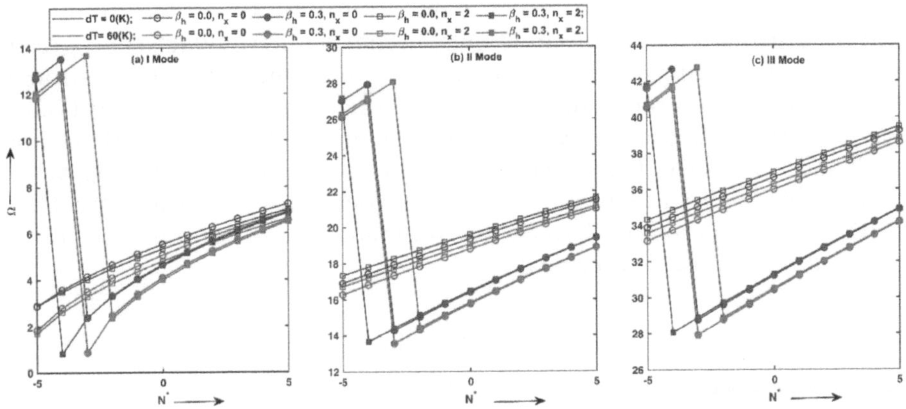

FIGURE 3.13
The frequency parameter Ω versus in-plane load parameter N^* for BD-FG nonuniform S-S nanobeam with $L/h_0 = 20$, $\mu = 1$ and $n_z = 1$.

three boundary conditions and modes. The value of the frequency parameter Ω increases with increasing value of the in-plane load parameter, N^*. The rate of increment is higher for the S-S beam and found in the order of S-S>C-S>C-C. This effect is more pronounced for a higher value of dT and β_h, while decreases with increasing values of n_x. The rate of increase decreases with an increase in the mode number. It is also clear that the C-S and S-S nanobeams with $\beta_h = 0.3$, $n_x = 2$, $dT = 60(\text{K})$ beam stop vibrating for higher compressive load, i.e., $N^* > -1$.

The value of the critical buckling load parameter N_{cr}^* for different values of beam parameters have been reported in Table 3.2. It has been shown that the critical buckling loads for C-C beams are higher than those for the corresponding C-S and S-S beams. The value of the critical buckling load decreases with the increase in the value of dT, β_h and μ. The critical buckling load decreases with increasing values of n_x for the C-S beam, while increases for the C-C beam and the S-S beam, keeping other parameters fixed. The effect increases with the increase in temperature for S-S and C-S boundary conditions, while gets reversed for C-C beams.

3.5 Summary

This chapter presents a comprehensive mathematical framework for analysing the vibrational behaviour of smart BD-FG tapered nanobeams subjected to simultaneous in-plane mechanical and thermal loads. The study demonstrates that transitioning the in-plane load from compressive to tensile enhances the effective stiffness of the nanobeam, thereby improving its thermal stability. It is observed that the critical buckling load diminishes with increasing temperature gradients, higher nonlocal effects, and greater geometric nonuniformity. Additionally, the influence of AFG is strongly dependent on both the nonuniformity parameter and the imposed boundary conditions. Importantly, incorporating material gradation along both axial and transverse directions—combined with geometric tapering, which enables more efficient and purpose-specific design of nanobeam structures, particularly in thermomechanical environments where performance optimization at the nanoscale is critical.

References

Ebrahimi, Farzad, and Mohammad Reza Barati. 2017. "Through-the-Length Temperature Distribution Effects on Thermal Vibration Analysis of Nonlocal Strain-Gradient Axially Graded Nanobeams Subjected to Nonuniform Magnetic Field." *Journal of Thermal Stresses* 40(5):548–63. doi: 10.1080/01495739.2016.1254076.

Ebrahimi, Farzad, Mohammad Reza Barati, and Parisa Haghi. 2017. "Thermal Effects on Wave Propagation Characteristics of Rotating Strain Gradient Temperature-Dependent Functionally Graded Nanoscale Beams." *Journal of Thermal Stresses* 40(5):535–47. doi: 10.1080/01495739.2016.1230483.

Khaniki, Hossein Bakhshi, and Sundaramoorthy Rajasekaran. 2018. "Mechanical Analysis of Non-Uniform Bi-Directional Functionally Graded Intelligent Micro-Beams Using Modified Couple Stress Theory." *Materials Research Express* 5(5). doi: 10.1088/2053-1591/aabe62.

Mahinzare, Mohammad, Mohammad Mostafa Barooti, and Majid Ghadiri. 2018. "Vibrational Investigation of the Spinning Bi-Dimensional Functionally Graded (2-FGM) Micro Plate Subjected to Thermal Load in Thermal Environment." *Microsystem Technologies* 24(3):1695–711. doi: 10.1007/s00542-017-3544-0.

Nemat-Alla, Mahmoud. 2003. "Reduction of Thermal Stresses by Developing Two-Dimensional Functionally Graded Materials." *International Journal of Solids and Structures* 40:7339–56. doi: 10.1016/j.ijsolstr.2003.08.017.

Palaniselvam, Thangavelu, and Jong Beom Baek. 2015. "Graphene Based 2D-Materials for Supercapacitors." *2D Materials* 2(3). doi: 10.1088/2053-1583/2/3/032002.

Rajasekaran, Sundaramoorthy, and Hossein Bakhshi. 2017. "Bending, Buckling and Vibration of Small-Scale Tapered Beams." *International Journal of Engineering Science* 120:172–88. doi: 10.1016/j.ijengsci.2017.08.005.

Samorì, Paolo, I. A. Kinloch, Xumeng Feng, and Vincenzo Palermo. 2015. "Graphene-Based Nanocomposites for Structural and Functional Applications: Using 2-Dimensional Materials in a 3-Dimensional World." *2D Materials* 2(3). doi: 10.1088/2053-1583/2/3/030205.

Shafiei, Navvab, Majid Ghadiri, and Mohammad Mahinzare. 2017. "Flapwise Bending Vibration Analysis of Rotary Tapered Functionally Graded Nanobeam in Thermal Environment." *Mechanics of Advanced Materials and Structures* 26(2):1–17. doi: 10.1080/15376494.2017.1365982.

Shafiei, Navvab, and Mohammad Kazemi. 2017. "Buckling Analysis on the Bi-Dimensional Functionally Graded Porous Tapered Nano-/Micro-Scale Beams." *Aerospace Science and Technology* 66:1–11. doi: 10.1016/j.ast.2017.02.019.

Shen, Zhi Bin, Dao Kui Li, Dong Li, and Guo Jin Tang. 2012. "Frequency Shift of a Nanomechanical Sensor Carrying a Nanoparticle Using Nonlocal Timoshenko Beam Theory." *Journal of Mechanical Science and Technology* 26(5):1577–83. doi: 10.1007/s12206-012-0338-2.

4

Influence of In-Plane Load and Nonlinear Thermal Field on the Vibration Response of Embedded BD-FG Nanobeams

4.1 Overview

The nanostructures resting on an elastic foundation have achieved great interest in embedded systems, particularly in MEMS/NEMSs and various body parts of modern aircraft, where they are operated in a multidirectional temperature field under the influence of in-plane load (Gopal et al. 2010; Hosseini et al. 2018). These embedded nanostructures help designers automatically control and isolate the vibrations, in order to avoid the situations which may cause failure of NEMSs and are concerned with the safety and comforts in aircraft (Fkb et al. 2012; Mathew, Joy, and George 2019). The interaction of structures and elastic media has been modelled using different models of the foundation, such as the Winkler foundation (Mohammadi et al. 2017; Rabhi et al. 2020; Wen, Tang, and Yang 2020), the Pasternak foundation (Murmu and Pradhan 2009), the Visco-Pasternak foundation (Ebrahimi and Hosseini 2016), and nonlinear elastic foundation (Abdullah et al. 2020). In this regard, a number of studies analysing the effect of elastic foundations on static/dynamic behaviour of uniform/non-uniform FG nanobeams subjected to mechanical and thermal load have appeared in the literature, and recent ones are given in references (Ebrahimi, Ehyaei, and Babaei 2016; Ebrahimi and Barati 2017; Mouffoki et al. 2017; Al-Furjan et al. 2020). However, no such work dealing with BD-FG nanobeams has been reported so far.

In the present chapter, a numerical model based on shear deformation theory in conjunction with Eringen's nonlocal theory has been presented to investigate the thermo-elastic vibration behaviour of BD-FG nonuniform thick nanobeams supported on the Pasternak foundation. The considered beam is subjected to NTD along the thickness and uniform mechanical in-plane loading. Two-directional variations in material properties are described by a power-law model across the thickness and as per the exponential function along the length. Employing Hamilton's energy principle,

DOI: 10.1201/9781003583806-4

the governing equations have been derived and then solved for the frequency parameter via GDQM for clamped, simply-supported, and a combination of both boundary conditions. A parametric study has been carried out to understand the vibration characteristics of the considered nanobeams, and it has been revealed that the combination of mechanical and thermal in-plane loads with the foundation parameter is the most critical combination for bi-directional functionally graded nanobeams. Results have been compared with the published work and found to be in good agreement.

4.2 Mathematical Framework

Referred to Cartesian coordinate system (x, y, z), consider a BD-FG nonuniform nanobeam of length L, width b, and variable thickness $h(x)$ subjected to a tensile, uniform in-plane load N_0 in a nonlinear thermal, resting on the Pasternak foundation. The middle plane being $z = 0$ with origin at the end $x = 0$. The top $(z = +h/2)$ and bottom $(z = -h/2)$ surfaces of the beam are fully ceramic and metallic, respectively. The nonuniformity in the cross-section of the beam arises due to thickness variation along the length as $h(x) = h_0\left(1 + \beta_h \frac{x}{L}\right)$, where β_h denotes the rate of the cross-sectional change along the thickness and h_0 is the thickness of the beam at the origin (Figure 4.1).

4.2.1 Energy Relations

In addition to Chapters 2 and 3, the effect of the elastic foundation is considered using the Pasternak's foundation model. Accordingly, the potential energy stored in the nanobeam due to shear and transverse reactions of the elastic foundation is

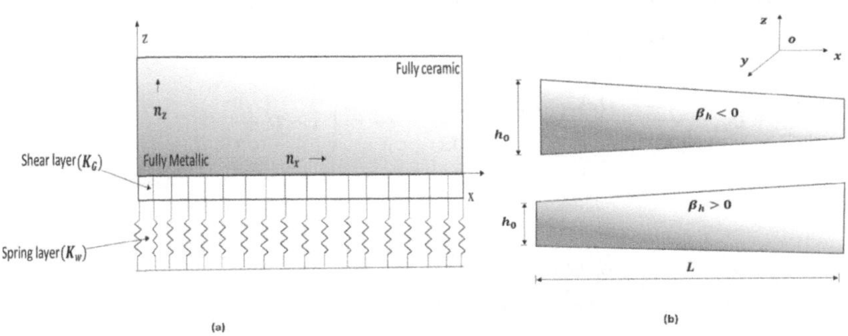

FIGURE 4.1
(a) Material distribution in BD-FG nanobeam resting on the Pasternak's foundation (b) Thickness variation of nanobeam in Cartesian-Coordinate system

$$U_f = \frac{1}{2}\int_0^b\int_0^L\left(K_w(w_0)^2 + K_G\left(\frac{\partial w_0}{\partial x}\right)^2\right)dxdy \tag{4.2.1}$$

where K_w and K_G are the transverse and shear moduli of the Pasternak's foundation.

Taking the variations of U_f and integrating with respect to y, gives

$$\delta U_f = \int_0^L b\left(K_w w_0(\delta w_0) + K_G\frac{\partial w_0}{\partial x}\delta\left(\frac{\partial w_0}{\partial x}\right)\right)dx \tag{4.2.2}$$

4.2.2 Nonlocal Governing Equations of Motion

Following the usual Hamilton procedure for $\delta L = \delta K - \delta U - \delta U^T - \delta U_{axial} - \delta U_f$ and substituting the energy expressions from equations (4.2.5), (4.2.7), (5.2.2) and (6.2.2), the Euler-Lagrange equations for the considered model are obtained as:

$$\frac{\partial N_{xx}^{nl}}{\partial x} = I_0\frac{\partial^2 u_0}{\partial t^2} + I_1\frac{\partial^2 \psi_x}{\partial t^2} \tag{4.2.3}$$

$$\frac{\partial M_{xx}^{nl}}{\partial x} - Q_{xz}^{nl} = I_1\frac{\partial^2 u_0}{\partial t^2} + I_2\frac{\partial^2 \psi_x}{\partial t^2} \tag{4.2.4}$$

$$\frac{\partial Q_{xz}^{nl}}{\partial x} + \left(N^T + N_0\right)\frac{\partial^2 w_0}{\partial x^2} - K_w w_0 + K_G\frac{\partial^2 w_0}{\partial x^2} = I_0\frac{\partial^2 w_0}{\partial t^2} \tag{4.2.5}$$

Following Eringen's nonlocal theory and using the expression for nonlocal stresses and moments (2.2.9), the nonlocal governing equations of motion (4.2.3)–(4.2.5) are given by

$$A_{xx}\frac{\partial^2 u_0}{\partial x^2} + B_{xx}\frac{\partial^2 \psi_x}{\partial x^2} + \frac{\partial A_{xx}}{\partial x}\frac{\partial u_0}{\partial x} + \frac{\partial B_{xx}}{\partial x}\frac{\partial \psi_x}{\partial x} = \left(1 - \mu\frac{\partial^2}{\partial x^2}\right)\left(I_0\frac{\partial^2 u_0}{\partial t^2} + I_1\frac{\partial^2 \psi_x}{\partial t^2}\right) \tag{4.2.6}$$

$$B_{xx}\frac{\partial^2 u_0}{\partial x^2} + D_{xx}\frac{\partial^2 \psi_x}{\partial x^2} + \frac{\partial B_{xx}}{\partial x}\frac{\partial u_0}{\partial x} + \frac{\partial D_{xx}}{\partial x}\frac{\partial \psi_x}{\partial x} - C_{xz}\left(\frac{\partial w_0}{\partial x} + \psi_x\right) = \left(1 - \mu\frac{\partial^2}{\partial x^2}\right)\left(I_1\frac{\partial^2 u_0}{\partial t^2} + I_2\frac{\partial^2 \psi_x}{\partial t^2}\right) \tag{4.2.7}$$

$$C_{xz}\frac{\partial^2 w_0}{\partial x^2}+C_{xz}\frac{\partial \psi_x}{\partial x}+\frac{\partial C_{xz}}{\partial x}\left(\frac{\partial w_0}{\partial x}+\psi_x\right)$$

$$+\left(1-\mu\frac{\partial^2}{\partial x^2}\right)\left\{\frac{\partial}{\partial x}\left(\left(N^T+N_0\right)\frac{\partial w_0}{\partial x}\right)-K_w w_0+K_G\frac{\partial^2 w_0}{\partial x^2}\right\} \tag{4.2.8}$$

$$=\left(1-\mu\frac{\partial^2}{\partial x^2}\right)\left(I_0\frac{\partial^2 w_0}{\partial t^2}\right)$$

For harmonic vibration, assume that $u_0(x,t)=\tilde{u}_0 e^{i\omega t}$, $w_0(x,t)=\tilde{w}_0 e^{i\omega t}$, $\psi_x(x,t)=\tilde{\psi}_x e^{i\omega t}$ and using non-dimensional variables $\bar{w}_0=\frac{\tilde{w}_0}{L}$, $\bar{u}_0=\frac{\tilde{u}_0}{L}$, $\bar{\psi}_x=\tilde{\psi}_x$, $X=\frac{x}{L}$, $H=\frac{h_0}{L}$ and $Z=\frac{z}{h}$, equations (4.2.6)–(4.2.8) reduce as follows:

$$U_0\frac{d^2\bar{u}_0}{dX^2}+U_1\frac{d^2\bar{\psi}_x}{dX^2}+\frac{dU_0}{dX}\frac{d\bar{u}_0}{dX}+\frac{dU_1}{dX}\frac{d\bar{\psi}_x}{dX}=-\hat{\Omega}^2\left(1-\gamma^2\frac{d^2}{dX^2}\right)\left[V_0\bar{u}_0+V_1\bar{\psi}_x\right] \tag{4.2.9}$$

$$U_1\frac{d^2\bar{u}_0}{dX^2}+U_2\frac{d^2\bar{\psi}_x}{dX^2}+\frac{dU_1}{dX}\frac{d\bar{u}_0}{dX}+\frac{dU_2}{dX}\frac{d\bar{\psi}_x}{dX}-U_3\left(\frac{d\bar{w}_0}{dX}+\bar{\psi}_x\right)$$

$$=-\hat{\Omega}^2\left(1-\gamma^2\frac{d^2}{dX^2}\right)\left[V_1\bar{u}_0+V_2\bar{\psi}_x\right] \tag{4.2.10}$$

$$U_3\frac{d^2\bar{w}_0}{dX^2}+U_3\frac{d\bar{\psi}_x}{dX}+\frac{dU_3}{dX}\left(\frac{d\bar{w}_0}{dX}+\bar{\psi}_x\right)$$

$$+\left(1-\gamma^2\frac{d^2}{dX^2}\right)\left\{\left(\frac{d}{dX}\left(P_1\frac{d\bar{w}_0}{dX}\right)\right)+\frac{H^3}{12}\left(N^*\frac{d^2\bar{w}_0}{dX^2}\right)-\frac{H^3}{12}\left(\bar{K}_w\bar{w}_0-\bar{K}_G\frac{\partial^2\bar{w}_0}{\partial x^2}\right)\right\}$$

$$=-\hat{\Omega}^2\left(1-\gamma^2\frac{d^2}{dX^2}\right)\left[V_0\bar{w}_0\right] \tag{4.2.11}$$

where

$$U_0=e^{n_x X}\left(1+\beta_h X\right)\bar{A}_{xx}H,\quad U_1=e^{n_x X}\left(1+\beta_h X\right)^2\bar{B}_{xx}H^2,$$

$$U_2=e^{n_x X}\left(1+\beta_h X\right)^3\bar{D}_{xx}H^3,\quad U_3=e^{n_x X}\left(1+\beta_h X\right)\bar{C}_{xz}H,$$

$$V_0 = e^{n_x X}(1+\beta_h X)\bar{I}_0 H, \quad V_1 = e^{n_x X}(1+\beta_h X)^2 \bar{I}_1 H^2,$$

$$V_2 = e^{n_x X}(1+\beta_h X)^3 \bar{I}_2 H^3; \quad P_1 = e^{2n_x X}(1+\beta_h X) H \alpha_0^c \bar{N}^T,$$

$$\bar{N}^T = \frac{N^T}{bhE_0^c}, N^* = \frac{NL^2}{\bar{D}_{xx}}, \bar{K}_w = \frac{K_w L^4}{\bar{D}_{xx}}, \bar{K}_G = \frac{K_G L^2}{\bar{D}_{xx}}$$

$$\hat{\Omega}^2 = \frac{\omega^2 L^2 \rho_0^c}{E_0^c}, (\bar{A}_{xx}, \bar{B}_{xx}, \bar{D}_{xx}) = \left(\frac{A_{xx}}{bhE_0^c}, \frac{B_{xx}}{bh^2 E_0^c}, \frac{D_{xx}}{bh^3 E_0^c}\right),$$

$$\bar{C}_{xz} = \frac{C_{xz}}{bhE_0^c} \text{ and } (\bar{I}_0, \bar{I}_1, \bar{I}_2) = \left(\frac{I_0}{bh\rho_0^c}, \frac{I_1}{bh^2\rho_0^c}, \frac{I_2}{bh^3\rho_0^c}\right).$$

The above governing equations are coupled differential equations with variable coefficients. It is solved numerically using GDQM and validated with published work.

4.3 Numerical Procedure

Now, employing GDQM, the discretization of governing equations (4.2.9)–(4.2.11) at internal grid points $(j = 2, 3, \ldots, (n-1))$ and C-C boundary condition (6.2.12) gives a complete set of $3n$ equations in $3n$ unknowns, which can be written in matrix form as

$$\begin{bmatrix} [\boldsymbol{K}_{dd}] & [\boldsymbol{K}_{db}] \\ [\boldsymbol{K}_{bd}^{CC}] & [\boldsymbol{K}_{bb}^{CC}] \end{bmatrix} \begin{Bmatrix} \{\boldsymbol{F}_d\} \\ \{\boldsymbol{F}_b\} \end{Bmatrix} - \hat{\Omega}^2 \begin{bmatrix} [\boldsymbol{M}_{dd}] & [\boldsymbol{M}_{db}] \\ [\boldsymbol{M}_{bd}^{CC}] & [\boldsymbol{M}_{bb}^{CC}] \end{bmatrix} \begin{Bmatrix} \{\boldsymbol{F}_d\} \\ \{\boldsymbol{F}_b\} \end{Bmatrix} = [0] \tag{4.3.1}$$

Similarly, for C-S and S-S boundary conditions, the governing equations are

$$\begin{bmatrix} [\boldsymbol{K}_{dd}] & [\boldsymbol{K}_{db}] \\ [\boldsymbol{K}_{bd}^{CS}] & [\boldsymbol{K}_{bb}^{CS}] \end{bmatrix} \begin{Bmatrix} \{\boldsymbol{F}_d\} \\ \{\boldsymbol{F}_b\} \end{Bmatrix} - \hat{\Omega}^2 \begin{bmatrix} [\boldsymbol{M}_{dd}] & [\boldsymbol{M}_{db}] \\ [\boldsymbol{M}_{bd}^{CS}] & [\boldsymbol{M}_{bb}^{CS}] \end{bmatrix} \begin{Bmatrix} \{\boldsymbol{F}_d\} \\ \{\boldsymbol{F}_b\} \end{Bmatrix} = [0] \tag{4.3.2}$$

$$\begin{bmatrix} [\boldsymbol{K}_{dd}] & [\boldsymbol{K}_{db}] \\ [\boldsymbol{K}_{bd}^{SS}] & [\boldsymbol{K}_{bb}^{SS}] \end{bmatrix} \begin{Bmatrix} \{\boldsymbol{F}_d\} \\ \{\boldsymbol{F}_b\} \end{Bmatrix} - \hat{\Omega}^2 \begin{bmatrix} [\boldsymbol{M}_{dd}] & [\boldsymbol{M}_{db}] \\ [\boldsymbol{M}_{bd}^{SS}] & [\boldsymbol{M}_{bb}^{SS}] \end{bmatrix} \begin{Bmatrix} \{\boldsymbol{F}_d\} \\ \{\boldsymbol{F}_b\} \end{Bmatrix} = 0 \tag{4.3.3}$$

4.4 Results and Discussion

The developed formulation of the present work obtained in equations (4.3.1)–(4.3.3) has been used to obtain the frequency parameter $\Omega = \left(\frac{\sqrt{12}L}{h_0}\right)\hat{\Omega}$, for the first three modes. In addition to the beam's parameter in Chapter 3, the foundation parameters is taken as: $\bar{K}_G = 0\ (1)\ 10$, $\bar{K}_W = 0\ (10)\ 100$.

The numerical results of the frequency parameter Ω have been given in Figures 4.2–4.4 for the first mode of vibration, as the behaviour in the second and third modes is almost the same. However, for selected data, the values of Ω have also been reported in Tables 4.1 and 4.2. The result in Table 4.1 show that the value of Ω is found to increase significantly with the increasing

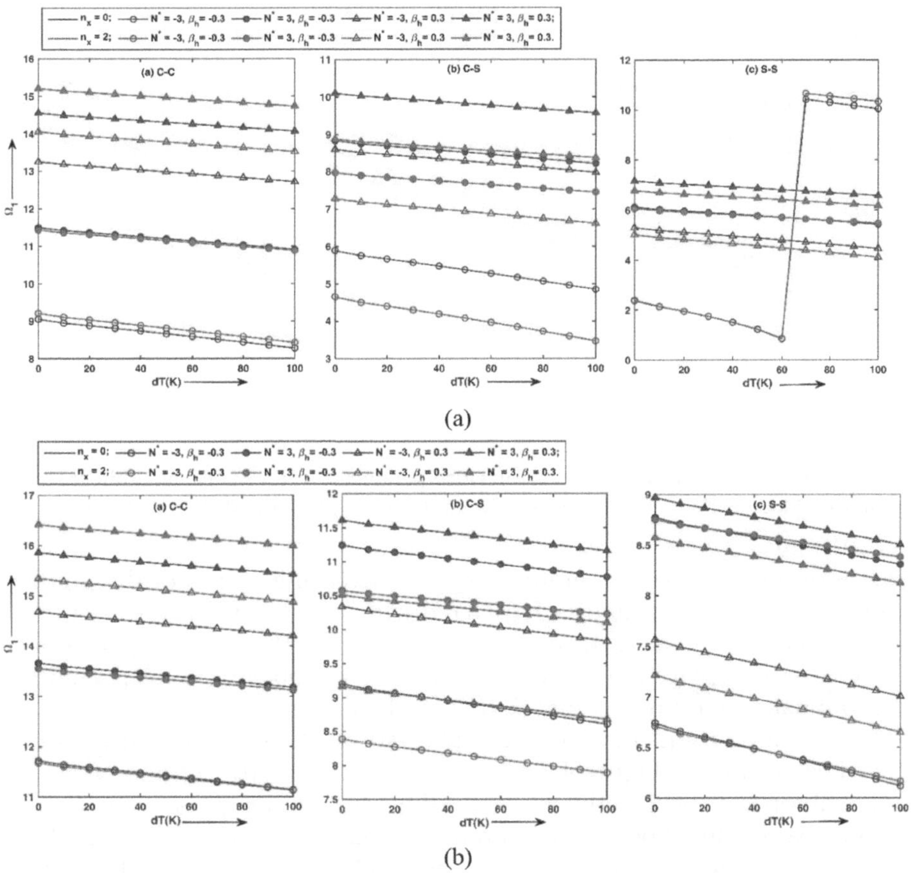

FIGURE 4.2
The frequency parameter Ω temperature difference dT for BD-FG nonuniform nanobeams in I-Mode with $\mu = 1$, $n_z = 1$, $L/h_0 = 20$: (a) $(\bar{K}_G, \bar{K}_w) = (0, 0)$ and (b) $(\bar{K}_G, \bar{K}_w) = (5, 25)$.

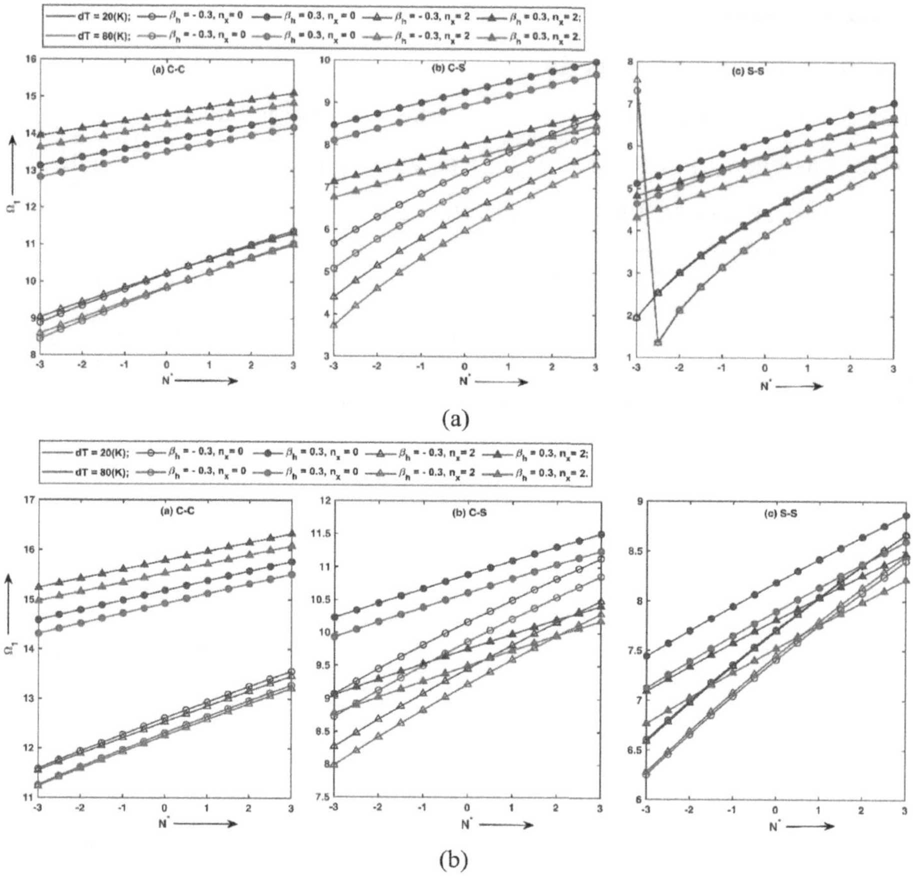

FIGURE 4.3
The frequency parameter Ω versus in-plane load parameter N^* for BD-FG nonuniform nanobeams in I-Mode with $\mu = 1$, $n_z = 1$, $L/h_0 = 20$: (a) $\left(\bar{K}_G, \bar{K}_w\right) = (0, 0)$ and (b) $\left(\bar{K}_G, \bar{K}_w\right) = (5, 25)$.

values of the foundation parameters. The increasing values of the gradient index n_z, nonlocal parameter μ have a reducing effect on Ω. As the nonuniformity parameter increases from $\beta_h = -0.5$ to $\beta_h = 0.5$, the frequency parameter Ω increases except for the nonlocal C-S and S-S nanobeams in the presence of the foundation parameters for the first mode of vibration.

The graphs of frequency parameter Ω versus the temperature difference dT have been drawn in Figure 4.2, taking. $n_x = 0, 2$, $\beta_h = -0.3, 0.3$ and $N^* = -3, 3$, in the absence as well as the presence of the foundation parameter. The frequency parameter Ω decreases with the increasing value of dT, for all three boundary conditions. This effect decreases in the presence of an elastic foundation, i.e., $(\bar{K}_G, \bar{K}_w) = (5, 25)$. The rate of decrease is higher for the beam thin

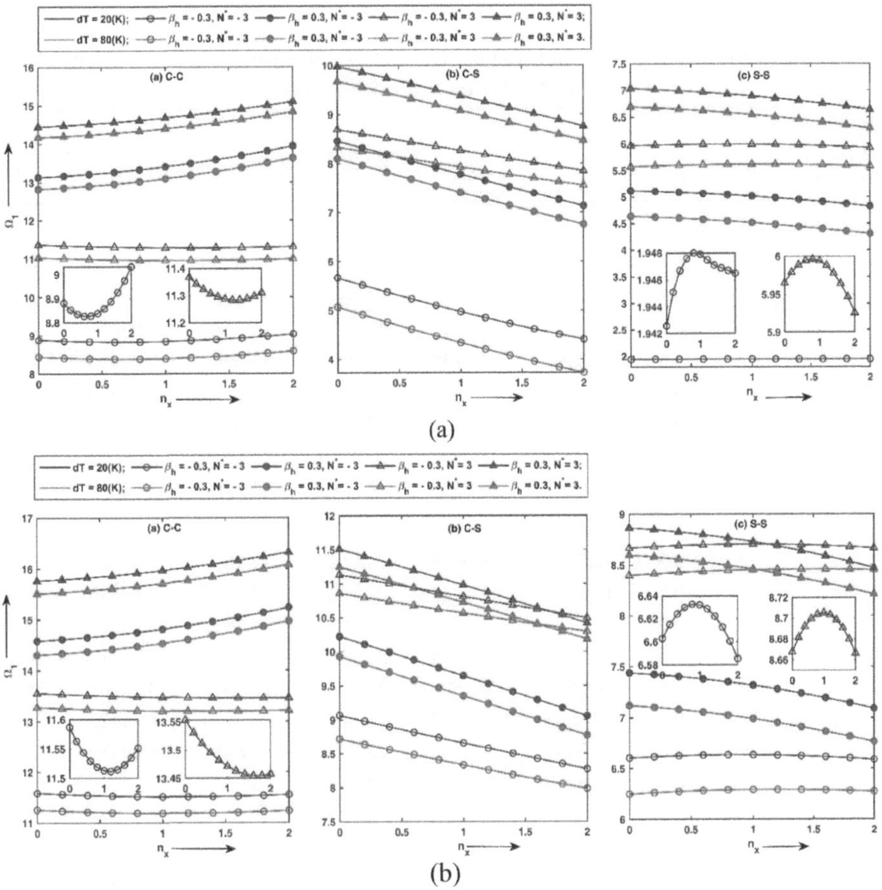

FIGURE 4.4
The frequency parameter Ω versus AFG n_x for BD-FG nonuniform nanobeams in I-Mode with $\mu = 1$, $n_z = 1$, $L/h_0 = 20$: (a) $\left(\bar{K}_G, \bar{K}_w\right) = (0, 0)$ and (b) $\left(\bar{K}_G, \bar{K}_w\right) = (5, 25)$.

at the outer edge ($\beta_h = -0.3$) with a higher value of AFG and subjected to compressive in-plane load. This effect is more noticeable on the S-S beam and found in the order of S-S>C-S>C-C boundary conditions.

Figure 4.3 shows the curves for the frequency parameter Ω versus N^* taking $n_x = 0, 2$, $\beta_h = -0.3, 0.3$, $dT = 20, 80(\text{K})$ and $\left(\bar{K}_G, \bar{K}_w\right) = (0,0), (5, 25)$ in the first mode. It is clear that the frequency parameter Ω increases as the in-plane load changes from compressive to tensile load in all modes, keeping other parameters fixed. This effect increases as the beam becomes thinner and thinner towards the outer edge. Considering axial gradation in the material properties of the beam, i.e., $n_x > 0$ also has a remarkable effect on the behaviour of Ω with N^*. For the nanobeam embedded on the foundation, the rate

TABLE 4.1

The Value of Frequency Parameter Ω for the BD-FG Nonuniform Nanobeam Having $n_x = 2$, $L/h_0 = 20$, $dT = 40(\text{K})$ for All Three Boundary Conditions

N^*			−3			3		
$(\bar{K}_G, \bar{K}_w)$	n_z	β_h	Ω_1	Ω_2	Ω_3	Ω_1	Ω_2	Ω_3
					C-C			
(0,0)	0	−0.5	10.4037	24.9065	39.9958	16.3350	34.8610	53.6717
		0.5	24.7644	51.6321	64.3859	26.4384	54.4943	81.5881
	5	−0.5	3.6843	9.7510	16.0955	16.3350	34.8610	53.6717
		0.5	11.5292	24.1138	36.2475	12.6991	26.1096	39.0268
(5,50)	0	−0.5	17.6192	34.5017	52.4620	21.5679	41.6960	62.4466
		0.5	26.9442	54.3912	81.1815	28.5166	57.1159	84.9727
	5	−0.5	9.0294	17.0381	25.6601	11.5058	21.6493	32.0873
		0.5	13.0447	26.0383	38.7464	14.1119	27.8960	41.3350
					C-S			
(0,0)	0	−0.5	3.5290	19.0488	34.0697	12.1858	29.2525	47.3695
		0.5	12.6852	41.1744	67.9663	12.1858	29.2525	47.3695
	5	−0.5	*	*	*	6.2843	14.6260	23.4530
		0.5	5.7443	19.1984	31.7535	8.1921	17.2710	26.4552
(5,50)	0	−0.5	14.1815	29.1500	46.3381	17.9992	36.0701	55.8471
		0.5	15.6883	43.8872	71.3202	17.4159	46.4426	74.8265
	5	−0.5	7.5443	14.5665	22.7809	9.8414	18.9419	28.8634
		0.5	7.8193	21.0887	34.0969	8.9596	22.8277	36.4908
					S-S			
(0,0)	0	−0.5	*	*	*	9.0731	24.7116	42.2778
		0.5	9.0606	34.9550	61.0122	11.1292	37.4586	64.3900
	5	−0.5	*	*	*	4.7452	12.4002	20.9547
		0.5	3.9805	16.2625	28.5011	5.4476	18.0108	30.8594
(5,50)	0	−0.5	11.7544	24.9215	41.4756	14.9316	31.1542	50.1649
		0.5	12.5213	37.5938	64.1474	14.1579	39.9590	67.3728
	5	−0.5	6.3979	12.5451	20.4347	8.2905	16.4613	25.9818
		0.5	6.3662	18.1019	30.6913	7.4302	19.7113	32.8956

of increase in Ω with N^* decreases. This rate of increase is higher in the presence of a high-temperature field. The effect of N^* decreases with the increase in the number of modes. The increment rate is higher for the S-S beam and found in the order of S-S>C-S>C-C boundary conditions. It is worth noting that the S-S nanobeam with $\beta_h = -0.3$, $dT = 80(\text{K})$ stops vibrating for higher value of the compressive load, i.e., $N^* < -1$.

Figure 4.4 demonstrates the effect of AFG n_x on frequency parameter Ω taking $N^* = -3, 3$, $\beta_h = -0.3, 0.3$, $dT = 20, 80(\text{K})$ and $(\bar{K}_G, \bar{K}_w) = (0,0), (5,25)$.

TABLE 4.2

The Effect of n_x on Ω of BD-FG Nanobeam with $\beta_h = -0.5$ for C-C and C-S Beams

		S-S				C-C			
		$N^* = -3$		$N^* = 3$		$N^* = -3$		$N^* = 3$	
$(\bar{K}_G, \bar{K}_w)$	Ω_i	$n_x = 0$	$n_x = 2$	$n_x = 0$	$n_x = 2$	$n_x = 0$	$n_x = 2$	$n_x = 0$	$n_x = 2$
(0,0)	1	3.3422	3.6843	8.3227	8.1921	*	*	4.5404	4.7452
	2	9.7265	9.7510	17.4490	17.2710	*	*	12.2994	12.4002
	3	16.2357	16.0955	26.6038	26.4552	*	*	20.8950	20.9547
(5,50)	1	9.1256	9.0294	11.5475	11.5058	6.1994	6.3979	8.0437	8.2905
	2	17.2085	17.0381	21.8357	21.6493	12.4675	12.5451	16.3401	16.4613
	3	25.7999	25.6601	32.2586	32.0873	20.3845	20.4347	25.9300	25.9818

It has been found that the effect of n_x on Ω depends on the nonuniformity parameter, boundary conditions, and modes. In the case of the C-S beam, the frequency parameter Ω decreases monotonically with the increasing value of n_x, for all other beam parameters. The rate of decay is more noticeable for the beam that are thicker at the outer edge, subjected to a compressive in-plane load, in the absence as well as the presence of a foundation.

However, for C-C and S-S nanobeams with $\beta_h \geq 0$, the frequency parameter Ω increases continuously with the increasing values of n_x in all modes except for the S-S beam vibrating in the first mode, where it is just the reverse. The rate of change decreases as the beam becomes thinner and thinner towards the outer edge. In particular, for the nanobeams with $\beta_h \leq -0.3$, there exists a point of minima in the vicinity of $n_x = 1$ for C-C and S-S beams, except for the S-S beam in the first mode, where maxima exist in the vicinity of $n_x = 1$ and for $\beta_h = -0.5$ the behaviour is just the reverse for both the boundary conditions. The corresponding results for the beam with $n_z = 1$, $\mu = 1$, $L / h_0 = 20$, $dT = 80(\mathrm{K})$ have been given in Table 4.2. The increase in the foundation parameters contributes to the stiffness of the embedded beam, which decreases the influence of the in-plane load, temperature difference, and nonuniformity parameter on the frequency parameter Ω as n_x varies. For S-S and C-S beams, the point of minima shifts towards the higher values of n_x except for the first mode of S-S beam where maxima shift towards the lower values of n_x as $\bar{K}_G, \bar{K}_w$ and N^* increases.

4.5 Summary

In summary, this chapter highlights the role of considering the effect of the Pasternak's foundation in the optimized designing of smart nanostructures subjected to thermal and in-plane loads. This analysis reveals that the presence of a foundation adds to the stiffness of the embedded nanobeam and increases its capacity to withstand in-plane and thermal load. This effect is more noticeable when the beam is subjected to compressive load and graded axially. These models help address key challenges, such as the interaction between nanostructures and the surrounding elastic medium through Pasternak's foundation models, and the complexities involved in nanoscale material design.

References

Abdullah, Sardar S., Shahrokh Hosseini-Hashemi, Nazhad A. Hussein, and Reza Nazemnezhad. 2020. "Thermal Stress and Magnetic Effects on Nonlinear Vibration of Nanobeams Embedded in Nonlinear Elastic Medium." *Journal of Thermal Stresses* 43(10):1316–32. doi: 10.1080/01495739.2020.1780175.

Al-Furjan, M. S. H., Hamed Safarpour, Mostafa Habibi, Mehran Safarpour, and Abdelouahed Tounsi. 2020. "A Comprehensive Computational Approach for Nonlinear Thermal Instability of the Electrically FG-GPLRC Disk Based on GDQ Method." *Engineering with Computers* (0123456789). doi: 10.1007/s00366-020-01088-7.

Ebrahimi, Farzad, and Mohammad Reza Barati. 2017. "Buckling Analysis of Nonlocal Third-Order Shear Deformable Functionally Graded Piezoelectric Nanobeams Embedded in Elastic Medium." *Journal of the Brazilian Society of Mechanical Sciences and Engineering* 39(3):937–52. doi: 10.1007/s40430-016-0551-5.

Ebrahimi, Farzad, Javad Ehyaei, and Ramin Babaei. 2016. "Thermal Buckling of FGM Nanoplates Subjected to Linear and Nonlinear Varying Loads on Pasternak Foundation." *Advances in Materials Research* 5(4):245–61.

Ebrahimi, Farzad, and S. H. S. Hosseini. 2016. "Thermal Effects on Nonlinear Vibration Behaviour of Viscoelastic Nanosize Plates." *Journal of Thermal Stresses* 39(5):606–25. doi: 10.1080/01495739.2016.1160684.

Fkb, A. G. E. et al. 2012. "Micro-and Nano-Scaled Sensors and Acuators for Aerospace Applications." *2012 Ninth International Conference on Networked Sensing (INSS)* 51–54.

Gopal, Hara, Mani Pakala, Ibrahim Khan, and K. V. S. V. N. Raju. 2010. "Sensors Integration in Embedded Systems." *2010 International Conference on Power, Control and Embedded Systems* 15–19.

Hosseini, M., M. R. Mofidi, A. Jamalpoor, and M. Safi Jahanshahi. 2018. "Nanoscale Mass Nanosensor Based on the Vibration Analysis of Embedded Magneto-Electro-Elastic Nanoplate Made of FGMs via Nonlocal Mindlin Plate Theory." *Microsystem Technologies* 24(5):2295–316. doi: 10.1007/s00542-017-3654-8.

Mathew, Jinu, Josny Joy, and Soney C. George. 2019. "Potential Applications of Nanotechnology in Transportation: A Review." *Journal of King Saud University - Science* 31(4):586–94. doi: 10.1016/j.jksus.2018.03.015.

Mohammadi, Kianoosh, Mohammad Mahinzare, Ali Rajabpour, and Majid Ghadiri. 2017. "Comparison of Modeling a Conical Nanotube Resting on the Winkler

Elastic Foundation Based on the Modified Couple Stress." *The European Physical Journal Plus* 132:115. doi: 10.1140/epjp/i2017-11395-x.

Mouffoki, Abderrahmane, E. A. Adda Bedia, Mohammed Sid Ahmed Houari, Abdelouahed Tounsi, and S. R. Mahmoud. 2017. "Vibration Analysis of Nonlocal Advanced Nanobeams in Hygro-Thermal Environment Using a New Two-Unknown Trigonometric Shear Deformation Beam Theory." *Smart Structures and Systems* 20(3):369–83. doi: 10.12989/sss.2017.20.3.369.

Murmu, T., and S. C. Pradhan. 2009. "Vibration Analysis of Nano-Single-Layered Graphene Sheets Embedded in Elastic Medium Based on Nonlocal Elasticity Theory." *Journal of Applied Physics* 105(6). doi: 10.1063/1.3091292.

Rabhi, Mohamed et al. 2020. "A New Innovative 3-Unknowns Hsdt for Buckling and Free Vibration of Exponentially Graded Sandwich Plates Resting on Elastic Foundations under Various Boundary Conditions." *Geomechanics and Engineering* 22(2):119–32. doi: 10.12989/gae.2020.22.2.119.

Wen, Chengqian, Lei Tang, and Guotao Yang. 2020. "Buckling and Post-Buckling of Pinned Euler Beams on Weakened Winkler Foundation under Thermal Loading." *Journal of Thermal Stresses* 43(5):529–42. doi: 10.1080/01495739.2020.1734128.

5

Influence of Surface Energies on Nonlocal Vibrations of BD-FG Nanobeams

5.1 Overview

The characteristic of large surface to volume ratio of composite nanostructures makes them more suitable for use in nanodevices such as capacitors (Chang et al. 2015), surgical devices (Liu et al. 2008), sensors (Calzzani et al. 2008), and highly sensitive detectors (Malani, Chaudhari, and Sambhe 2015). Keeping this in view, researchers and engineers have focused on investigating the effect of surface energies on the mechanical behaviour of composite nanostructures (Moniem et al. 2013; Farajpour, Dehghany, and Shahidi 2013; Hosseini and Rahmani 2016; Lei et al. 2012; Li, Zhang, and Yong 2014; Preethi, Rajagopal, and Narasimha 2015; Streitz, Cammarata, and Sieradzki 1994; Wang and Wang 2014), to mention a few. Among these, most of the studies are based on the surface-elasticity model proposed by Gurtin and Murdoch (Gurtin and Ian Murdoch 1975; Murdoch 2005). Further, it has been observed that the surface effect strongly influences the nonlocal stress and deformation properties of nanostructures (Li, Lin, and Ng 2020). In view of this, numerous studies analysing the combined effect of nonlocal and surface stresses on the static/dynamic behaviour of FG nanostructures have appeared in the literature, and the recent work has been reported in references (Ansari et al. 2015; Chen, Sun, and Li 2017; Yang et al. 2017). Among these, Hosseini-Hashemi, Nazemnezhad, and Bedroud (2014) obtained the analytical solutions for nonlinear frequencies of FG simply-supported Euler–Bernoulli nanobeams under the influence of surface effect using Eringen's nonlocal theory. Ansari and Norouzzadeh (2016) exhibited static buckling responses of circular, elliptical, and skew FG nanoplates considering the nonlocal and surface effects. Based on Eringen's nonlocal theory, MCST, and surface elasticity theory (SET), Attia (2017) presented a model to study the buckling (Naskar et al. 2022) and vibration characteristics of an FG Euler–Bernoulli nanobeam. Ebrahimi (2017) explored the free vibration of an FG Euler-Bernoulli nanobeam with surface layers resting on a viscoelastic foundation in the thermal field based on NSGT and SET. Arefi et al. (2019) studied the vibration behaviour of the FG Timoshenko nanobeam resting on a nonlinear foundation for various

DOI: 10.1201/9781003583806-5

distributions of material properties using NSGT and SET by considering the effect of dielectricity, piezoelectricity, and flexoelectricity. Ebrahimi and Heidari (2019) explored the nonlinear free vibrations of FG rectangular Reddy nanoplates embedded in an elastic medium based on Eringen's nonlocal theory and G-M SET. Very recently, the effect of various thermal, electric, and magnetic fields has been investigated on the mechanical behaviour of BD-FG nanobeams/nanoplates using Eringen's nonlocal theory in conjunction with SET (Mirjavadi et al. 2017; Shafiei and Kazemi 2017; Yang et al. 2018) and NSGT (Karami, Janghorban, and Rabczuk 2019; Rahmani, Faroughi, and Friswell 2020; Sahmani and Safaei 2019) without involving surface effect.

In this chapter, the effect of surface energy on the vibration behaviour of the BD-FG nanobeam with surface layers has been studied based on Timoshenko beam theory, together with G-M SET and ENT. The surface effect has been incorporated employing the G-M SET, while the nonlocal effect is being captured using Eringen's nonlocal theory. The material properties of the beam are taken to be graded according to a power-law model in the thickness and axial directions. Using Hamilton's energy principle, the governing equations and corresponding boundary conditions have been obtained for the present model. These governing equations have been solved for the frequency parameter using GDQM for the first three modes and three different combinations of clamped and simply-supported boundary conditions. A parametric study has been presented for elaborating the influence of surface parameters, axial and transverse gradient indices, the nonlocal parameter, slenderness ratio, and modes on the frequency parameter for the three boundary conditions.

5.2 Mathematical Formulation

Consider a BD-FG nanobeam of length L, breadth b, and thickness h in the Cartesian coordinate system (x, y, z). The nanobeam is assumed to be composed of surface layers and a bulk part, as shown in Figure 5.1. The top $(z = +h/2)$ and bottom $(z = -h/2)$ surfaces of the beam are fully ceramic and metallic, respectively.

5.2.1 Material Properties of BD-FG Nanobeam with Surface Layers

The BD-FG nanobeam has been tailored by mixing metallic and ceramic materials. The material properties are taken to be graded in the thickness and length directions according to a power-law distribution. Accordingly, the effective material property P_f of the BD-FG nanobeam is taken as (Karami, Janghorban, and Rabczuk 2019)

$$P_f(x, z) = \left[(P_c - P_m) \left(x/L \right)^{n_x} \left(\frac{z}{h} + \frac{1}{2} \right)^{n_z} + P_m \right] \tag{5.2.1}$$

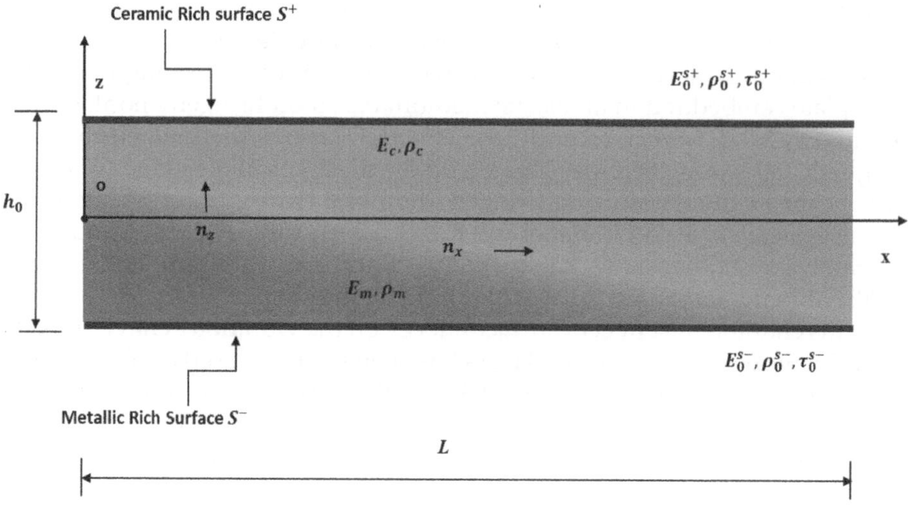

FIGURE 5.1
Distribution of material properties of the BD-FG nanobeam with surface layers in the Cartesian coordinate system.

5.2.2 Strain-Displacement Relations

According to HOSDT (Moniem et al. 2014), the displacement components at any point of the nanobeam are given by,

$$u_x(x,z,t) = -z\frac{\partial w_0(x,t)}{\partial x} + R(z)\left(\frac{\partial w_0(x,t)}{\partial x} + \psi_x(x,t)\right),$$
$$u_y(x,z,t) = 0, \quad u_z(x,t) = w_0(x,t) \tag{5.2.2}$$

where w_0 is the transverse displacement of the mid-plane, ψ_x is the rotational angle of yz-plane about the transverse direction. The function $R(z)$ is a shape function describing the distribution of transverse shear stresses through the thickness of the beam. The appropriate form of shape function for EBT and TBT are as follows:

i. $R(z) = 0$ for EBT
ii. $R(z) = z$ for TBT

The strain components are

$$\varepsilon_{xx} = \frac{\partial u_x}{\partial x} = (R - z)\frac{\partial^2 w_0}{\partial x^2} + R\frac{\partial \psi_x}{\partial x}, \quad \gamma_{xz} = \frac{\partial u_x}{\partial z} + \frac{\partial u_z}{\partial x} = \frac{dR}{dz}\left(\frac{\partial w_0}{\partial x} + \psi_x\right) \tag{5.2.3}$$

5.2.3 Surface Elasticity Theory

The surface elasticity theory proposed by Gurtin-Murdoch is the most widely used and efficient continuum theory to incorporate the effect of surface energy on the mechanical response of nanostructures. According to the G-M SET (Attia 2017), a solid is assumed to be composed of a bulk part and surface layers. The surface layers are simulated as a membrane of negligible thickness with different material properties from the underlying bulk part, which are perfectly bonded. The constitutive stress–strain relations for the two surface layers are

$$t_{xx}^{s\pm} = \tau_0^{s\pm} \pm \left(\lambda_0^{s\pm} + 2\mu_0^{s\pm}\right)\varepsilon_{xx}^{s\pm}, \quad t_{xz}^{s\pm} = \tau_0^{s\pm}\frac{\partial w_0(x,t)}{\partial x} \tag{5.2.4}$$

where $\tau_0^{s\pm}$ are residual surface tensions. $\lambda_0^{s\pm}$, $2\mu_0^{s\pm}$ are the surface Lamé constants and $\varepsilon_{xx}^{s\pm}$ are strain components of the surfaces $S^{\pm}$ at $z = \pm h/2$.

In the classical continuum theories, the normal stress component t_{zz} has not been considered because t_{zz} is very small in comparison to other stress components. However, in the case of surface effect, this doesn't satisfy the equilibrium conditions at surfaces $S^{\pm}$. Hence, classical continuum theories have been modified by assuming the linear variation of t_{zz} through the thickness of the beam, which satisfies the equilibrium conditions (Ansari et al. 2015) in the two surfaces. Accordingly, the classical normal stress t_{zz} is given by (Hosseini-Hashemi, Nazemnezhad, and Bedroud 2014),

$$\begin{aligned} t_{zz} = {} & \frac{1}{2}\left[\left(\frac{\partial t_{xz}^{s+}}{\partial x} - \frac{\partial t_{xz}^{s-}}{\partial x}\right) - \left(\rho^{s+}\frac{\partial^2 w_0}{\partial t^2} - \rho^{s-}\frac{\partial^2 w_0}{\partial t^2}\right)\right] \\ & + \frac{z}{h}\left[\left(\frac{\partial t_{xz}^{s+}}{\partial x} + \frac{\partial t_{xz}^{s-}}{\partial x}\right) - \left(\rho^{s+}\frac{\partial^2 w_0}{\partial t^2} + \rho^{s-}\frac{\partial^2 w_0}{\partial t^2}\right)\right] \end{aligned} \tag{5.2.5}$$

Using equation (5.2.4), normal stress t_{zz} is obtained as

$$\begin{aligned} t_{zz} = {} & \frac{1}{2}\left[\left(\tau_0^{s+} - \tau_0^{s-}\right)\frac{\partial^2 w_0}{\partial x^2} - \left(\rho^{s+}\frac{\partial^2 w_0}{\partial t^2} - \rho^{s-}\frac{\partial^2 w_0}{\partial t^2}\right)\right] \\ & + \frac{z}{h}\left[\left(\tau_0^{s+} + \tau_0^{s-}\right)\frac{\partial^2 w_0}{\partial x^2} - \left(\rho^{s+}\frac{\partial^2 w_0}{\partial t^2} + \rho^{s-}\frac{\partial^2 w_0}{\partial t^2}\right)\right] \end{aligned} \tag{5.2.6}$$

Therefore, the classical constitutive relation for the normal stress t_{xx} in bulk will be modified as

$$t_{xx} = \frac{E(x,z)}{\left(1-\nu^2\right)}\varepsilon_{xx} + \frac{\nu}{(1-\nu)}t_{zz} \tag{5.2.7}$$

5.2.4 Nonlocal Stress–Strain Relations and Stress Resultants and Moments

The modified form of the nonlocal stress–strain relation (2.2.8) for the bulk part is

$$\sigma_{xx}^{nl}-\mu\frac{\partial^2\sigma_{xx}^{nl}}{\partial x^2}=\frac{E(x,z)}{\left(1-\nu^2\right)}\varepsilon_{xx}+\frac{\nu}{(1-\nu)}t_{zz},\quad \sigma_{xz}^{nl}-\mu\frac{\partial^2\sigma_{xz}^{nl}}{\partial x^2}=t_{xz}=G(x,z)\gamma_{xz} \tag{5.2.8}$$

where $G(=E/(2(1+\nu))$ is the shear modulus of elasticity and the constant ν is Poisson's ratio, taken to be 0.3 in this study.

The nonlocal stress–strain relation for the surface layers $S^{\pm}$ is

$$\sigma_{xx}^{nl(S\pm)}-\mu\frac{\partial^2\sigma_{xx}^{nl(S\pm)}}{\partial x^2}=\tau_0^{s\pm}\pm\left(\lambda_0^{s\pm}+2\mu_0^{s\pm}\right)\varepsilon_{xx}^{s\pm} \tag{5.2.9}$$

$$\sigma_{xz}^{nl\,(S\pm)}-\mu\frac{\partial^2\sigma_{xz}^{nl(S\pm)}}{\partial x^2}=\left(\tau_0^{s\pm}\right)\frac{\partial w_0}{\partial x} \tag{5.2.10}$$

The nonlocal stress resultants and moments for bulk and surface layers are defined as follows:

$$\left(M_{xx}^{nl},Q_{xz}^{nl}\right)=\int_{-b/2-h/2}^{-b/2}\int^{h/2}\left(z\sigma_{xx}^{nl},k_s\sigma_{xz}^{nl}\right)dydz \tag{5.2.11}$$

$$\left(M_{xx}^{nl(S+)},M_{xx}^{nl(S-)},Q_{xz}^{nl(S+)},Q_{xz}^{nl(S-)}\right)=\left(zb\sigma_{xx}^{nl(S+)}\Big|_{\frac{h}{2}},\ zb\sigma_{xx}^{nl(S-)}\Big|_{-\frac{h}{2}},bk_s\sigma_{xz}^{nl(S+)},bk_s\sigma_{xz}^{nl(S-)}\right) \tag{5.2.12}$$

5.2.5 Energy Relations

The first variation of strain energy stored in the BD-FG nanobeam with a surface layer is

$$\delta U=\int_v\left(z\sigma_{xx}^{nl}\delta\varepsilon_{xx}+\sigma_{xz}^{nl}\delta\gamma_{xz}\right)dv+\oint_{S^+}\sigma_{xx}^{nl(S+)}\delta\varepsilon_{xx}^{s+}+\sigma_{xz}^{nl(S+)}\delta\varepsilon_{xz}^{s+}ds+\oint\sigma_{xx}^{nl(S-)}\delta\varepsilon_{xx}^{s-}+\sigma_{xz}^{nl(S-)}\delta\varepsilon_{xz}^{s-}ds \tag{5.2.13}$$

The variation of the kinetic energy of the BD-FG beam with surface layers

$$\delta K = \int_v \rho\left(\frac{\partial u_x}{\partial t}\frac{\partial \delta u_x}{\partial t} + \frac{\partial u_z}{\partial t}\frac{\partial \delta u_z}{\partial t}\right)dv + \oint_{S^+} \rho^{S+}\left(\frac{\partial u_x}{\partial t}\frac{\partial \delta u_x}{\partial t} + \frac{\partial u_z}{\partial t}\frac{\partial \delta u_z}{\partial t}\right)dS +$$

$$\oint_{S^-} \rho^{S-}\left(\frac{\partial u_x}{\partial t}\frac{\partial \delta u_x}{\partial t} + \frac{\partial u_z}{\partial t}\frac{\partial \delta u_z}{\partial t}\right)dS \tag{5.2.14}$$

5.2.6 Nonlocal Governing Equations of Motion Based on Timoshenko's Beam Theory

For developing the governing equations of motion, following the usual Hamilton's energy principle for $\delta L = \delta K - \delta U$, Euler-Lagrange equations for the present nanobeam model are obtained as:

$$\frac{\partial M_{xx}^{nl}}{\partial x} + \frac{\partial}{\partial x}\left(M_{xx}^{nl(S+)} + M_{xx}^{nl(S-)}\right) - Q_{xz}^{nl} = I_2 \frac{\partial^2 \psi_x}{\partial t^2} \tag{5.2.15}$$

$$\frac{\partial Q_x^{nl}}{\partial x} + \frac{\partial}{\partial x}\left(Q_{xz}^{nl(S+)} + Q_{xz}^{nl(S-)}\right) = I_0 \frac{\partial^2 w_0}{\partial t^2} \tag{5.2.16}$$

where mass moments of inertia $\left(I_0(x), I_2(x)\right)$ are given by

$$\left(I_0(x), I_2(x)\right) = \int_{-h/2}^{h/2} b\rho(x,z)\left(1, z^2\right)dz + \left(b\left(\rho^{S+} + \rho^{S-}\right), \frac{h^2 b\left(\rho^{S+} + \rho^{S-}\right)}{4}\right)$$

For Timoshenko nanobeam taking $R(z) = z$, the nonlocal stress moments (equations 5.2.11 and 5.2.12) of the nanobeam with surface layers in terms of deflections are

$$\left(1 - \mu^2 \frac{\partial^2}{\partial x^2}\right)\left(M_{xx}^{nl} + M_{xx}^{nl(S+)} + M_{xx}^{nl(S-)}\right) = \left(A_1 + A_3\right)\frac{\partial \psi_x}{\partial x} + A_2 \frac{\partial^2 w_0}{\partial x^2} - S_1 \frac{\partial^2 w_0}{\partial t^2} + C_1 \tag{5.2.17}$$

$$\left(1 - \mu^2 \frac{\partial^2}{\partial x^2}\right)\left(Q_x^{nl} + Q_x^{nl(S+)} + Q_x^{nl(S-)}\right) = B_1\left(\frac{\partial w_0}{\partial x} + \psi_x\right) + B_2 \frac{\partial w_0}{\partial x} \tag{5.2.18}$$

where $A_1 = b\int_{-h/2}^{h/2} z^2 \frac{E(x,z)}{\left(1 - \nu^2\right)}dz$; $A_2 = b\int_{-h/2}^{h/2} \frac{z^2}{h}\frac{\nu}{(1-\nu)}\left(\tau_0^{s+} + \tau_0^{s-}\right)dz$;

$$A_3 = \frac{h^2 b\left(\lambda_0^{s+} + 2\mu_0^{s+} + \lambda_0^{s-} + 2\mu_0^{s-}\right)}{4}; B_1 = b \int_{-h/2}^{h/2} k_s G(x,z)dz; B_2 = \left(\tau_0^{s+} + \tau_0^{s-}\right);$$

$$S_1 = b \int_{-h/2}^{h/2} \frac{z^2}{h} \frac{\nu}{(1-\nu)} \left(\rho_0^{s+} + \rho_0^{s-}\right)dz; C_1 = \frac{h}{2}\left(\tau_0^{s+} - \tau_0^{s-}\right).$$

Similarly, by substituting the nonlocal stress resultants from equations (5.2.15) and (5.2.16) into equations (5.2.17) and (5.2.18), the nonlocal governing equations of motion for free vibration of BD-FG Timoshenko nanobeam with surface layers are obtained as follows:

$$\begin{aligned} &A_2 \frac{\partial^3 w_0}{\partial x^3} + (A_1 + A_3)\frac{\partial^2 \psi_x}{\partial x^2} + \frac{d(A_1 + A_3)}{dx}\frac{\partial \psi_x}{\partial x} - B_1\left(\frac{\partial w_0}{\partial x} + \psi_x\right) \\ &= \left(1 - \mu \frac{\partial^2}{\partial x^2}\right)\left(I_2 \frac{\partial^2 \psi_x}{\partial t^2}\right) + S_1 \frac{\partial^3 w_0}{\partial x \partial t^2} \end{aligned} \tag{5.2.19}$$

$$B_1\left(\frac{\partial^2 w_0}{\partial x^2} + \frac{\partial \psi_x}{\partial x}\right) + \frac{dB_1}{dx}\left(\frac{\partial w_0}{\partial x} + \psi_x\right) + B_2 \frac{\partial^2 w_0}{\partial x^2} = \left(1 - \mu \frac{\partial^2}{\partial x^2}\right)\left(I_0 \frac{\partial^2 w_0}{\partial t^2}\right) \tag{5.2.20}$$

For harmonic solution, we have $w_0 = e^{i\omega t}\bar{w}_0(x)$, $\psi_x = e^{i\omega t}\bar{\psi}_x(x)$ and the governing equations are reduced as follows

$$\begin{aligned} &A_2 \frac{d^3 \bar{w}_0}{dx^3} + (A_1 + A_3)\frac{d^2 \bar{\psi}_x}{dx^2} + \frac{d(A_1 + A_3)}{dx}\frac{\bar{\psi}_x}{dx} - B_1\left(\frac{d\bar{w}_0}{dx} + \bar{\psi}_x\right) \\ &= -\omega^2\left[\left(1 - \mu \frac{d^2}{dx^2}\right)(I_2 \bar{\psi}_x) + S_1 \frac{d\bar{w}_0}{dx}\right] \end{aligned} \tag{5.2.21}$$

$$B_1\left(\frac{d^2 \bar{w}_0}{dx^2} + \frac{d\bar{\psi}_x}{dx}\right) + \frac{dB_1}{dx}\left(\frac{d\bar{w}_0}{dx} + \bar{\psi}_x\right) + B_2 \frac{d^2 \bar{w}_0}{dx^2} = -\omega^2\left(1 - \mu \frac{d^2}{dx^2}\right)(I_0 \bar{w}_0) \tag{5.2.22}$$

with boundary conditions:

i. Clamped-Clamped (C-C) beam:

$$\bar{w}_0 = \frac{d\bar{w}_0}{dx} = 0 \quad \text{at} \quad x = 0, L \tag{5.2.23}$$

ii. Clamped-Simply supported (C-S) beam:

$$\bar{w}_0 = \frac{d\bar{w}_0}{dx} = 0 \quad \text{at} \quad x = 0 \quad \text{and} \quad \bar{w}_0 = M_{xx}^{nl} + M_{xx}^{nl(S+)} + M_{xx}^{nl(S-)} = 0 \quad \text{at} \quad x = L \tag{5.2.24}$$

iii. Simply supported-Simply supported (S-S) beam

$$\bar{w}_0 = M_{xx}^{nl} + M_{xx}^{nl(S+)} + M_{xx}^{nl(S-)} = 0 \quad \text{at} \quad x = 0, L \tag{5.2.25}$$

The governing equations for Timoshenko (5.2.21) and (5.2.22) BD-FG nanobeams are differential equations with variable coefficients.

5.3 Solution Procedure

Employing GDQM, the discretized form of the governing equations (5.2.21) and (5.2.22) at internal grid points $(j = 2, 3, \ldots, (n-1))$ together with the C-C boundary condition (5.2.23) gives a complete set of n equations in n, which can be written in matrix form as

$$\begin{bmatrix} [\mathbf{K}_{dd}] & [\mathbf{K}_{db}] \\ [\mathbf{K}_{bd}^{CC}] & [\mathbf{K}_{bb}^{CC}] \end{bmatrix} \begin{Bmatrix} \{\mathbf{F}_d\} \\ \{\mathbf{F}_b\} \end{Bmatrix} - \hat{\Omega}^2 \begin{bmatrix} [\mathbf{M}_{dd}] & [\mathbf{M}_{db}] \\ [\mathbf{M}_{bd}^{CC}] & [\mathbf{M}_{bb}^{CC}] \end{bmatrix} \begin{Bmatrix} \{\mathbf{F}_d\} \\ \{\mathbf{F}_b\} \end{Bmatrix} = [0] \tag{5.3.1}$$

where $[\mathbf{K}_{dd}]$, $[\mathbf{M}_{dd}]$ and $[\mathbf{K}_{db}]$,$[\mathbf{M}_{db}]$ are the stiffness and mass matrices of order $2(n-2)\times 2(n-2)$ and $2(n-2)\times 4$, respectively. The matrices $[\mathbf{K}_{bd}^{CC}]$, $[\mathbf{M}_{bd}^{CC}]$ and $[\mathbf{K}_{bb}^{CC}]$, $[\mathbf{M}_{bb}^{CC}]$ are stiffness and mass matrices corresponding to C-C boundary conditions of order $4\times 2(n-2)$ and 4×4, respectively. The solution vector $\{\{\mathbf{F}_d\}\{\mathbf{F}_b\}\}^T$ is of the order $2n\times 1$ given by,

$$\begin{Bmatrix} \{\mathbf{F}_d\} \\ \{\mathbf{F}_b\} \end{Bmatrix} = \left\{ \left\{ \bar{w}_{02}, \bar{\psi}_{x2}, \ldots, \bar{w}_{0i}, \bar{\psi}_{xi}, \ldots, \bar{w}_{0(n-1)}, \bar{\psi}_{x(n-1)} \right\}^T \left\{ \bar{w}_{01}, \bar{\psi}_{x1}, \bar{w}_{0n}, \bar{\psi}_{xn} \right\}^T \right\}$$

where $\bar{w}_{0i}, \bar{\psi}_{xi}$'s denote $\bar{w}_0(X_i), \bar{\psi}_x(X_i)$, for $(i = 1, 2, \ldots, n)$.

Similarly, for C-S and S-S boundary conditions, the governing equations are

$$\left\{\begin{bmatrix}[\mathbf{K}_{dd}] & [\mathbf{K}_{db}]\\ [\mathbf{K}_{bd}^{CS}] & [\mathbf{K}_{bb}^{CS}]\end{bmatrix}-\hat{\Omega}^2\begin{bmatrix}[\mathbf{M}_{dd}] & [\mathbf{M}_{db}]\\ [\mathbf{M}_{bd}^{CS}] & [\mathbf{M}_{bb}^{CS}]\end{bmatrix}\right\}\begin{Bmatrix}\{\mathbf{F}_d\}\\ \{\mathbf{F}_b\}\end{Bmatrix}=[0] \tag{5.3.2}$$

$$\left\{\begin{bmatrix}[\mathbf{K}_{dd}] & [\mathbf{K}_{db}]\\ [\mathbf{K}_{bd}^{SS}] & [\mathbf{K}_{bb}^{SS}]\end{bmatrix}-\hat{\Omega}^2\begin{bmatrix}[\mathbf{M}_{dd}] & [\mathbf{M}_{db}]\\ [\mathbf{M}_{bd}^{SS}] & [\mathbf{M}_{bb}^{SS}]\end{bmatrix}\right\}\begin{Bmatrix}\{\mathbf{F}_d\}\\ \{\mathbf{F}_b\}\end{Bmatrix}=[0] \tag{5.3.3}$$

5.4 Results and Discussion

The generated systems of algebraic equations obtained for Timoshenko nanobeams (5.3.1)–(5.3.3) have been solved for the frequency parameter using MATLAB software. The lowest three eigenvalues are taken as the frequency parameter for the first three modes of vibration. In order to analyse the combined effect of surface and the nonlocal energies on vibration behaviour of BD-FG nanobeam, the results have been given for non-dimensional frequency parameter $\Omega\left(=\omega\frac{L^2}{h}\sqrt{\frac{12(1-\nu^2)\rho_0^c}{E_0^c}}\right)$. In the present paper, ceramic and metallic constituents of the nanobeam are silicon and aluminium, respectively. Following (Ebrahimi and Heidari 2019; Fan et al. 2020) their material properties for bulk and surface layers are given in Table 5.1. The values of other beam parameters are taken as:

Nonlocal parameter $\mu = 0, 1, \ldots, 2\ \text{nm}^2$.
Transverse functionally gradient index: $n_z = 0, 1, \ldots, 10$,
Axially functionally gradient index: $n_x = 0, 1, \ldots, 5$,
Slenderness ratio: $10 \le L/h_0 \le 50$.

TABLE 5.1

Material Properties of the BD-FG Nanobeam Constituents for Bulk and Surface Layers

Material	E(GPa)	ρ(kg/m³)	ρ_0(kg/m²)	τ_0(N/m)	λ_0(N/m)	μ_0(N/m)
Al	70	2,700	5.46e-7	0.9108	6.842	−0.3755
Si	210	2,370	3.17e-7	0.6048	−4.488	−2.774

Non-dimensional frequency parameter Ω for varying values of the nonlovcal parameter, gradient indices and slenderness ratio with or without considering surface effect has been plotted in Figures 5.2–5.4 and Table 5.2.

The curves of the frequency parameter Ω versus the nonlocal parameter μ have been plotted in Figures 5.2–5.4, taking of $n_z = 0,5$ and $n_x = 0,3$ for all three modes and boundary conditions. The value of Ω decreases continuously with the increasing value of μ. The rate of decrease is higher in the absence of AFG and TFG. However, the consideration of surface effects increases this

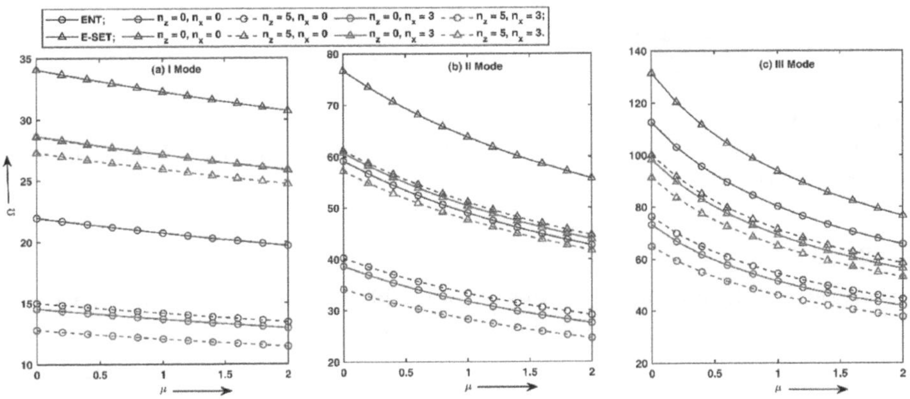

FIGURE 5.2
The frequency parameter Ω versus the nonlocal parameter μ for the BD-FG C-C nanobeam with surface layers having $L/h = 20$.

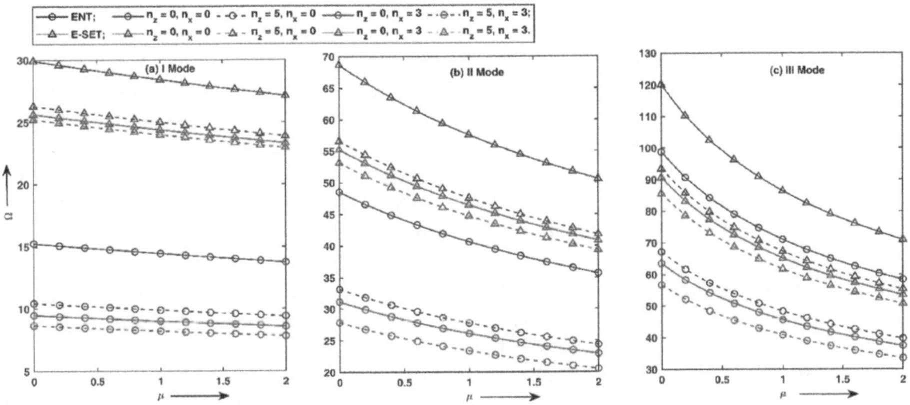

FIGURE 5.3
The frequency parameter Ω versus the nonlocal parameter μ for the BD-FG C-S nanobeam with surface layers having $L/h = 20$.

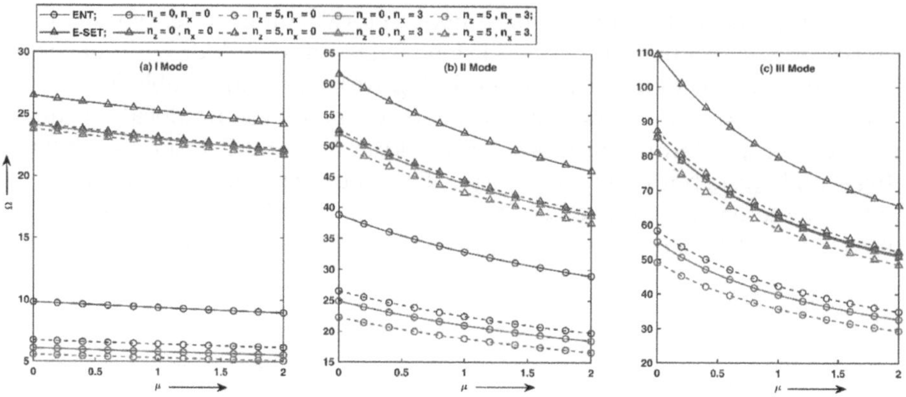

FIGURE 5.4
The frequency parameter Ω versus the nonlocal parameter μ for the BD-FG C-S nanobeam with surface layers having $L/h = 20$.

TABLE 5.2
The Values of the Frequency Parameter for the BD-FG Nanobeam with Surface Layers for All the Boundary Conditions

			Ω_1		Ω_2		Ω_3	
L/h_0	n_z	n_x	ENT	E-SET	ENT	E-SET	ENT	E-SET
					C-C			
50	1	0	16.6608	95.1205	40.2113	168.9854	67.5328	218.7407
		1	13.9853	92.4235	33.7820	163.9184	56.7243	211.4423
		10	12.1943	90.9422	28.9269	161.0595	48.2384	207.1488
	10	0	13.5417	91.5383	32.6837	162.3198	54.8912	209.3360
		1	12.4748	90.7159	30.1142	160.7749	50.5740	207.1082
		10	11.7652	90.4206	28.1661	160.1343	47.1463	206.0135
30	1	0	16.6532	47.4415	40.1428	86.0498	67.2888	115.3225
		1	13.9788	45.5211	33.7234	81.7675	56.5150	107.8759
		10	12.1886	44.5383	28.8767	79.2636	48.0619	103.1249
	10	1	13.5356	44.9858	32.6287	80.7109	54.6955	106.3406
		5	12.4691	44.3683	30.0630	79.3134	50.3916	103.8547
		10	11.7598	44.1235	28.1178	78.5656	46.9754	102.2578
					C-S			
50	1	0	11.5233	93.2925	32.9699	166.1043	58.9912	215.4970
		1	9.5098	90.7135	27.6571	161.2854	49.6272	208.6188
		10	7.9916	89.3047	23.1735	158.5838	41.6945	204.5662
	10	0	9.3661	90.3237	26.7978	160.4349	47.9484	207.2635
		1	8.5473	89.5281	24.6624	158.9460	44.1994	205.1326
		10	7.9346	89.2598	22.8335	158.3599	40.9597	204.1210

(Continued)

TABLE 5.2 (*Continued*)

The Values of the Frequency Parameter for the BD-FG Nanobeam with Surface Layers for All the Boundary Conditions

			Ω_1		Ω_2		Ω_3	
L/h_0	n_z	n_x	ENT	E-SET	ENT	E-SET	ENT	E-SET
30	1	0	11.5191	45.1491	32.9237	82.2874	58.8102	110.7723
		5	9.5064	43.4403	27.6181	78.5231	49.4723	104.2678
		10	7.9886	42.6055	23.1409	76.3230	41.5655	99.9942
	10	0	9.3626	43.3283	26.7607	78.0407	47.8032	103.2013
		5	8.5442	42.7675	24.6282	76.7902	44.0643	101.0058
		10	7.9317	42.5797	22.8015	76.1579	40.8338	99.5924
					S-S			
50	1	0	7.4324	91.5361	26.3730	163.3352	50.9575	212.3820
		1	6.2804	90.0291	22.2075	160.2493	42.8342	207.4915
		10	5.1378	88.6504	18.4741	157.6097	35.9474	203.5468
	10	0	6.0410	89.1420	21.4359	158.5995	41.4184	205.2454
		1	5.5733	88.8681	19.7610	157.9607	38.1674	204.0929
		10	5.1110	88.6070	18.2378	157.3913	35.3507	203.1130
30	1	0	7.4302	43.0726	26.3423	78.8604	50.8243	106.5933
		1	6.2786	42.2348	22.1813	76.4956	42.7202	101.6979
		10	5.1363	41.4698	18.4524	74.5041	35.8527	97.8794
	10	0	6.0392	41.7921	21.4112	75.5604	41.3116	100.2775
		1	5.5717	41.6153	19.7381	74.9221	38.0679	98.7764
		10	5.1095	41.4469	18.2165	74.3573	35.2582	97.5221

rate of decay. The effect of the nonlocal parameter on Ω is more noticeable on the C-C beam as compared to the corresponding C-S and S-S beams. This effect increases with the increase in the number of modes.

In order to extract the influence of the surface effect on the vibration characteristics of BD-FG nanobeam, the frequency parameter Ω with different values of AFG and TFG and slenderness ratio have been presented in Table 5.2, for the first three modes and boundary conditions. The consideration of the surface effect enlarges the frequency parameter. Overall, the surface effect plays an important role in determining the vibration characteristics of nanobeams and is more pronounced for higher values of AFG and TFG. For the case in hand, due to the consideration of surface effect, the value of the frequency parameter Ω increases by 893.71% or around 9 times from the corresponding one obtained without surface effect for isotropic $(n_z = 0, n_x = 0)$ S-S nanobeam with a fixed value of $\mu = 1$ and $L/h_0 = 50$ in the first mode. As the nanobeam changes its nature from isotropic to BD-FG, i.e., a change in (n_z, n_x) from $(0,0)$ to $(10, 5)$, this percentage increase in Ω is 1,593.16% or around 16 times. Further, it is also clear that the surface effect has a remarkable effect on the frequency parameter of the nanobeam with a large value of the slenderness ratio. This is due to the increase in the surface-to-volume ratio with the increasing value of L/h_0. The effect of surface is more pronounced for the S-S nanobeam in comparison with the C-S and S-S nanobeams. The increase in the number of modes decreases the surface effect.

5.5 Summary

In summary, this chapter developed the foundation of a mathematical framework for incorporating the effect of the nonlocal and surface energy on the vibration characteristics of BD-FG nanobeams. The chapter systematically discusses the various stages of deriving the nonlocal governing differential equations for vibrations of the BD-FG nanobeam with surface effect and their solution approach. The present analysis concludes that the consideration of surface effect strongly influences the stiffness of the nanobeam, due to the large surface-to-volume ratio of the nanobeam. Also, the bi-directional gradation of material properties amounts to the flexibility of the beam and makes BD-FG nanobeams more sensitive to surface parameters than the corresponding homogeneous/unidirectional FG nanobeams. The predictive framework developed here is generic and extended to a wide range of materials and geometry in the next Chapters 6 and 7. This makes it a valuable reference for future applications in materials science and engineering.

References

Ansari, Raza, M. A. Ashrafi, Talieh Pourashraf, and Saeid Sahmani. 2015. "Vibration and Buckling Characteristics of Functionally Graded Nanoplates Subjected to Thermal Loading Based on Surface Elasticity Theory." *Acta Astronautica* 109:42–51. doi: 10.1016/j.actaastro.2014.12.015.

Ansari, Ansari, and Amir Mohammad Norouzzadeh. 2016. "Nonlocal and Surface Effects on the Buckling Behaviour of Functionally Graded Nanoplates: An Isogeometric Analysis." *Physica E: Low-Dimensional Systems and Nanostructures* 84:84–97. doi: 10.1016/j.physe.2016.05.036.

Arefi, Mohammad, Mahmoud Pourjamshidian, Ali Ghorbanpour Arani, and Timon Rabczuk. 2019. "Influence of Flexoelectric, Small-Scale, Surface and Residual Stress on the Nonlinear Vibration of Sigmoid, Exponential and Power-Law FG Timoshenko Nano-Beams." *Journal of Low Frequency Noise, Vibration and Active Control* 38(1):122–42. doi: 10.1177/1461348418815410.

Attia, Mohamed A. 2017. "On the Mechanics of Functionally Graded Nanobeams with the Account of Surface Elasticity." *International Journal of Engineering Science* 115:73–101. doi: 10.1016/j.ijengsci.2017.03.011.

Calzzani, Jr., Fernando A., Redahegn Sileshi, Aschalew Kassu, Jean Michel Taguenang. 2008. "Detection of Residual Traces of Explosives by Surface Enhanced Raman Scattering Using Gold Coated Substrates Produced by Nanospheres Imprint Technique." *Optics and Photonics in Global Homeland Security IV* 6945(April 2008):694510. doi: 10.1117/12.777072.

Chang, Melina S., Taehoon Kim, Jong Hun Kang, J. Park, and Chong Rae Park. 2015. "The Effect of Surface Characteristics of Reduced Graphene Oxide on the Performance of a Pseudocapacitor." *2D Materials* 2(1). doi: 10.1088/2053-1583/2/1/014007.

Chen, D. Q., Dongliang Sun, and Xian Fang Li. 2017. "Surface Effects on Resonance Frequencies of Axially Functionally Graded Timoshenko Nanocantilevers with Attached Nanoparticle." *Composite Structures* 173:116–26. doi: 10.1016/j.compstruct.2017.04.006.

Ebrahimi, Farzad. 2017. "Vibration Analysis of Viscoelastic Inhomogeneous Nanobeams Incorporating Surface and Thermal Effects." *Applied Physics A* 123(1):1–10. doi: 10.1007/s00339-016-0511-z.

Ebrahimi, Farzad, and Ebrahim Heidari. 2019. "Surface Effects on Nonlinear Vibration of Embedded Functionally Graded Nanoplates via Higher Order Shear Deformation Plate Theory." *Mechanics of Advanced Materials and Structures* 26(8):671–99. doi: 10.1080/15376494.2017.1410908.

Fan, Fan, Biao Lei, Saeid Sahmani, and Babak Safaei. 2020. "On the Surface Elastic-Based Shear Buckling Characteristics of Functionally Graded Composite Skew Nanoplates." *Thin-Walled Structures* 154(February):106841. doi: 10.1016/j.tws.2020.106841.

Farajpour Ouderji, Ali, Mohammad Dehghany, and Alireza Shahidi. 2013. "Surface and Nonlocal Effects on the Axisymmetric Buckling of Circular Graphene Sheets in Thermal Environment." *Composites Part B: Engineering* 50:333–43. doi: 10.1016/j.compositesb.2013.02.026.

Gurtin, Morton E., and A. Ian Murdoch. 1975. "A Continuum Theory of Elastic Material Surfaces." *Archive for Rational Mechanics and Analysis* 57(4):291–323. doi: 10.1007/BF00261375.

Hosseini, Seyyed Amirhosein, and Omid Rahmani. 2016. "Surface Effects on Buckling of Double Nanobeam System Based on Nonlocal Timoshenko Model." *International Journal of Structural Stability and Dynamics* 16(10):1–17. doi: 10.1142/S0219455415500777.

Hosseini-Hashemi, Shahrokh, Reza Nazemnezhad, and Mohammad Bedroud. 2014. "Surface Effects on Nonlinear Free Vibration of Functionally Graded Nanobeams Using Nonlocal Elasticity." *Applied Mathematical Modelling* 38(14):3538–53. doi: 10.1016/j.apm.2013.11.068.

Karami, Behrouz, Maziar Janghorban, and Timon Rabczuk. 2019. "Dynamics of Two-Dimensional Functionally Graded Tapered Timoshenko Nanobeam in Thermal Environment Using Nonlocal Strain Gradient Theory." *Composites Part B: Engineering* 182(October 2019):107622. doi: 10.1016/j.compositesb.2019.107622.

Lei, Xiao Wen, Toshiaki Natsuki, Jin Xing Shi, and Qing Ni. 2012. "Surface Effects on the Vibrational Frequency of Double-Walled Carbon Nanotubes Using the Nonlocal Timoshenko Beam Model." *Composites Part B: Engineering* 43(1):64–69. doi: 10.1016/j.compositesb.2011.04.032.

Li, Li, Rongming Lin, and Teng Yong Ng. 2020. "Contribution of Nonlocality to Surface Elasticity." *International Journal of Engineering Science* 152(152):103311. doi: 10.1016/j.ijengsci.2020.103311.

Li, Xian-fang, Hua Zhang, and Kang Yong. 2014. "Dependence of Young's Modulus of Nanowires on Surface Effect." *International Journal of Mechanical Sciences* 81:120–25. doi: 10.1016/j.ijmecsci.2014.02.018.

Liu, Zhuang et al. 2008. "Drug Delivery with Carbon Nanotubes for in Vivo Cancer Treatment." *American Association for Cancer Research* 68(16):6652–60.

Malani, Akshata S., Anagha D. Chaudhari, and Rajeshkumar U. Sambhe. 2015. "A Review on Applications of Nanotechnology in Automotive Industry." *International Journal of Mechanical and Mechatronics Engineering* 10(1):36–40. doi: 10.5281/zenodo.1110867.

Mirjavadi, Seyed Sajad, Behzad Mohasel Afshari, Navvab Shafiei, A. M. S. Hamouda, and Mohammad Kazemi. 2017. "Thermal Vibration of Two-Dimensional Functionally Graded (2D-FG) Porous Timoshenko Nanobeams." *Steel and Composite Structures* 25(4):415–26.

Moniem, Mohammad Abdel, Mohamed Eltaher, A Khairy, Ayman Sadoun, and Fatema Alzahraa Omar. 2014. "Static and Buckling Analysis of Functionally Graded Timoshenko Nanobeams." *Applied Mathematics and Computation* 229:283–95. doi: 10.1016/j.amc.2013.12.072.

Moniem, Mohammad Abdel, Mohamed Eltaher, Fatin F Mahmoud, Amr Assie, and Efstathios Meletis. 2013. "Coupling Effects of Nonlocal and Surface Energy on Vibration Analysis of Nanobeams." *Applied Mathematics and Computation* 224:760–74. doi: 10.1016/j.amc.2013.09.002.

Murdoch, A. Ian. 2005. "Some Fundamental Aspects of Surface Modelling." *Journal of Elasticity* 80(1–3):33–52. doi: 10.1007/s10659-005-9024-2.

Naskar, Susmita, Kishor B. Shingare, Soumyadeep Mondal, and Tanmoy Mukhopadhyay. 2022. "Flexoelectricity and Surface Effects on Coupled Electromechanical Responses of Graphene Reinforced Functionally Graded Nanocomposites: A Unified Size-Dependent Semi-Analytical Framework." *Mechanical Systems and Signal Processing* 169(December 2021):108757. doi: 10.1016/j.ymssp.2021.108757.

Preethi, Kasirajan, Amirtham Rajagopal, and Junuthula Narasimha. 2015. "Surface and Non-Local Effects for Non-Linear Analysis of Timoshenko Beams." *International Journal of Non-Linear Mechanics* 76:100–11. doi: 10.1016/j.ijnonlinmec.2015.06.006.

Rahmani, Arash, Shirko Faroughi, and Michael Ian Friswell. 2020. "The Vibration of Two-Dimensional Imperfect Functionally Graded (2D-FG) Porous Rotating Nanobeams Based on General Nonlocal Theory." *Mechanical Systems and Signal Processing* 144:106854. doi: 10.1016/j.ymssp.2020.106854.

Sahmani, Saeid, and Babak Safaei. 2019. "Nonlinear Free Vibrations of Bi-Directional Functionally Graded Micro/Nano-Beams Including Nonlocal Stress and Microstructural Strain Gradient Size Effects." *Thin-Walled Structures* 140(November 2018):342–56. doi: 10.1016/j.tws.2019.03.045.

Shafiei, Navvab, and Mohammad Kazemi. 2017. "Buckling Analysis on the Bi-Dimensional Functionally Graded Porous Tapered Nano-/Micro-Scale Beams." *Aerospace Science and Technology* 66:1–11. doi: 10.1016/j.ast.2017.02.019.

Streitz, Frederick H., Robert C. Cammarata, and Karl Sieradzki. 1994. "Surface-Stress Effects on Elastic Properties. I. Thin Metal Films." *Physical Review B* 49(15):10699–706. doi: 10.1103/PhysRevB.49.10699.

Wang, Dong-Hui, and Gang-Feng Wang. 2014. "Surface Effects on the Vibration and Buckling of Double-Nanobeam-Systems Surface Effects on the Vibration and Buckling of Double-Nanobeam-Systems." *Journal of Nanomaterials*. doi: 10.1155/2011/518706.

Yang, Lihong, Tao Fan, Liping Yang, Xiao Han, and Zongbing Chen. 2017. "Bending of Functionally Graded Nanobeams Incorporating Surface Effects Based on Timoshenko Beam Model." *Theoretical and Applied Mechanics Letters* 7(3):152–58. doi: 10.1016/j.taml.2017.03.001.

Yang, Tianzhi, Ye Tang, Qian Li, and Xiao-dong Yang. 2018. "Nonlinear Bending, Buckling and Vibration of Bi-Directional Functionally Graded Nanobeams." *Composite Structures* 204(July):313–19. doi: 10.1016/j.compstruct.2018.07.045.

6

Effect of Nonuniform Geometry of Nanobeam on Nonlocal and Surface Energies of Nanobeam and Vibration Analysis

6.1 Overview

To enhance the functionality and design of smart multifunctional structures, engineers have increasingly adopted the strategy of introducing structural non-uniformity (Mukhopadhyay et al. 2020; Saumya, Naskar, and Mukhopadhyay 2023). In particular, BD-FG non-uniform nanostructures have gained significant attention recently, driven by the growing demand for long-term reliability and the optimal design of sensitive devices. Shafiei and Kazemi (2017) performed the nonlocal buckling analysis of two-directional FG tapered porous micro/nano beams. Nonlocal vibration characteristics of BD-FG nonuniform nanobeams subjected to a thermal field have been explored by Lal and Dangi (2019). A study on the buckling and vibration behaviour of an embedded BD-FG nonuniform tapered nanobeam has been presented by Lal and Dangi (2021b). Karami, Janghorban, and Rabczuk (2020) studied the dynamics of a BD-FG tapered Timoshenko nanobeam in the thermal environment using NSGT. The dynamic response of the nonuniform BD-FG nanobeam subjected to a thermal field has been presented by Despenić, Janevski, and Pavlović (2023) based on the refined beam theory and NSGT. Recent research on multi-directional functionally graded nanostructures taking surface stress into account has been published by R. Lal and C. Dangi, where the effect of surface energy on the nonlocal vibration behaviour of the BD-FG nanobeam has been studied using GM-SET coupled with ENT (Lal and Dangi 2021a) and NSGT (Dangi, Lal, and Sukavanam 2021). The investigation of the coupled nonlocal and surface effect has been primarily focused on a uniform structure up to this point.

This chapter applies the generalized mathematical model developed in Chapter 5 to fully characterize the vibrational behaviour of the BD-FG tapered nanobeams, incorporating surface effects. The study accounts for material property variations following a power-law distribution through the thickness and an exponential distribution along the beam's length. The mathematical framework from Chapter 5 is extended to accommodate the nonuniform

DOI: 10.1201/9781003583806-6

geometry of the nanobeam, and the GDMQ is employed to compute its natural frequencies under C–C and S–S boundary conditions. Furthermore, a comparative analysis is presented to examine the impact of nonlocal elasticity and surface stress effects on the vibrational response of nanobeams with linearly varying thickness, as opposed to those with uniform thickness.

6.2 Mathematical Formulation

Consider a BD-FG tapered nanobeam of length L, width b, and thickness $h(x)$ in a Cartesian coordinate system. The thickness of the nanobeam is assumed to vary linearly along the length as $h(x)=h_0\left(1+\alpha\frac{x}{L}\right)$, where α denotes the taper parameter, and h_0 represents the thickness of the nanobeam at the origin. The geometry and material composition of the considered nanobeam have been shown by highlighting its bulk part and surface layers. For the BD-FG nanobeam with surface layers, the upper surface $\left(S^+\right)$ of the beam at $z=h/2$ is ceramic-rich, while the bottom surface $\left(S^-\right)$ at $z=-h/2$ is metallic-rich. The beam is graded in the thickness direction as per the power-law model and as per the exponential law in the length direction. Accordingly, the specific material property $P(x,z)$ is as follows:

$$P(x,z)=e^{n_x\frac{n}{L}}\left[(P_c-P_m)\left(\frac{z}{h}+\frac{1}{2}\right)^{n_z}+P_m\right] \tag{6.2.1}$$

where indices m and c represent the metallic and ceramic constituents, respectively. The variation of elastic modulus and density of the nanobeam with the functionally gradient index and heterogeneity index has been shown in Figure 6.1.

6.2.1 Nonlocal Governing Equation of Motion

Using the mathematical framework of the governing differential equations for vibrations of the BD-FG nanobeams developed in Chapter 5, the nonlocal stress moments (equations 5.2.17 and 5.2.18) of a tapered nanobeam with surface layers in terms of deflections are

$$\left(1-\mu^2\frac{\partial^2}{\partial x^2}\right)M_{xx}=A_1\frac{\partial\psi_x}{\partial x}+A_2\frac{\partial^2 w_0}{\partial x^2}-S_2\frac{\partial^2 w_0}{\partial t^2}+C \tag{6.2.2}$$

$$\left(1-\mu^2\frac{\partial^2}{\partial x^2}\right)Q_{xz}=B_1\left(\frac{\partial w_0}{\partial x}+\psi_x\right);\quad\left(1-\mu^2\frac{\partial^2}{\partial x^2}\right)Q_x^S=S_1\frac{\partial w_0}{\partial x} \tag{6.2.3}$$

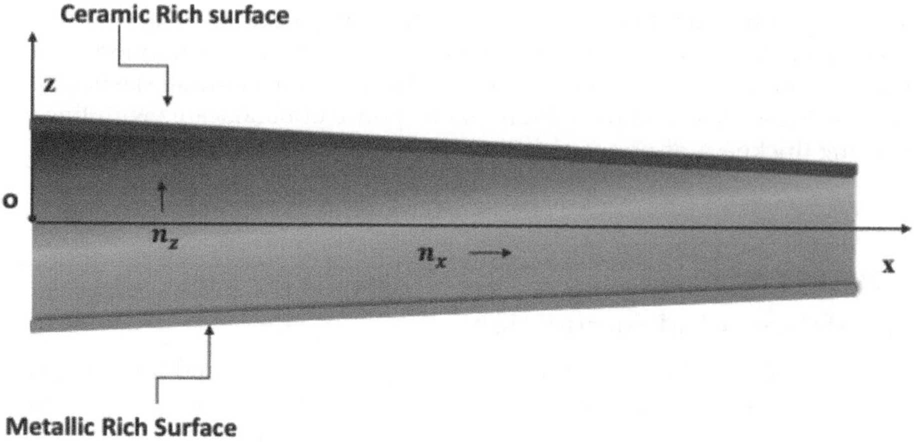

FIGURE 6.1
Geometry and material composition of the BD-FG tapered nanobeam with surface layers.

where, $A_1 = b\int_{-h(x)/2}^{h(x)/2} z^2 \frac{E(x,z)}{(1-\nu^2)}dz + \frac{h(x)^2 b\left(\lambda_0^{s+} + 2\mu_0^{s+} + \lambda_0^{s-} + 2\mu_0^{s-}\right)}{4}$;

$$A_2 = b\int_{-h(x)/2}^{h(x)/2} \frac{z^2}{h}\frac{\nu}{(1-\nu)}\left(\tau_0^{s+} + \tau_0^{s-}\right)dz;\ B_1 = b\int_{-h(x)/2}^{h(x)/2} k_s G(x,z)dz;\ S_1 = \left(\tau_0^{s+} + \tau_0^{s-}\right);$$

$$S_2 = b\int_{-h(x)/2}^{h(x)/2} \frac{z^2}{h}\frac{\nu}{(1-\nu)}\left(\rho_0^{s+} + \rho_0^{s-}\right)dz;\ C = \frac{h(x)}{2}\left(\tau_0^{s+} - \tau_0^{s-}\right).$$

Substituting the nonlocal stresses from equations (5.2.15) and (5.2.16) into equations (6.2.2) and (6.2.3), we obtain the nonlocal governing equations of motion as follows:

$$A_2(x)\frac{\partial^3 w_0}{\partial x^3} + \frac{dA_2(x)}{dx}\frac{\partial^2 w_0}{\partial x^2} + A_1(x)\frac{\partial^2 \psi_x}{\partial x^2} + \frac{dA_1(x)}{dx}\frac{\partial \psi_x}{\partial x} - B_1(x)\left(\frac{\partial w_0}{\partial x} + \psi_x\right)$$
$$= \left(1 - \mu\frac{\partial^2}{\partial x^2}\right)\left(I_2(x)\frac{\partial^2 \psi_x}{\partial t^2}\right) + S_2(x)\frac{\partial^3 w_0}{\partial x \partial t^2} + \frac{dS_2(x)}{dx}\frac{\partial^2 w_0}{\partial t^2} \quad (6.2.4)$$

$$\frac{d}{dx}B_1(x)\left(\frac{\partial w_0}{\partial x} + \psi_x\right) + S_1\frac{\partial^2 w_0}{\partial x^2} = \left(1 - \mu\frac{\partial^2}{\partial x^2}\right)\left(I_0(x)\frac{\partial^2 w_0}{\partial t^2}\right) \quad (6.2.5)$$

For harmonic vibrations, we have, $w_0 = e^{i\omega t}\bar{w}_0(x), \psi_x = e^{i\omega t}\bar{\psi}_x(x)$

$$A_2\frac{d^3\bar{w}_0}{dx^3}+\frac{dA_2(x)}{dx}\frac{d^2\bar{w}_0}{dx^2}+A_1\frac{d^2\bar{\psi}_x}{dx^2}+\frac{dA_1}{dx}\frac{\bar{\psi}_x}{dx}-B_1\left(\frac{d\bar{w}_0}{dx}+\bar{\psi}_x\right)$$
$$=-\omega^2\left[\left(1-\mu\frac{d^2}{dx^2}\right)(I_2\bar{\psi}_x)+S_2\frac{d\bar{w}_0}{dx}+\frac{dS_2(x)}{dx}\right] \tag{6.2.6}$$

$$B_1\left(\frac{d^2\bar{w}_0}{dx^2}+\frac{d\psi_x}{dx}\right)+\frac{dB_1}{dx}\left(\frac{d\bar{w}_0}{dx}+\bar{\psi}_x\right)+S_1\frac{d^2\bar{w}_0}{dx^2}=-\omega^2\left(1-\mu\frac{d^2}{dx^2}\right)(I_0\bar{w}_0) \tag{6.2.7}$$

The governing equations (6.2.6) and (6.2.7) for the BD-FG tapered nanobeams are differential equations with variable coefficients.

6.3 Solution Procedure

Employing GDQM, the discretized form of the governing equations (6.2.6) and (6.2.7) at internal grid points $(j=2,3,\ldots,(n-1))$ together with the C-C boundary condition (5.2.23) gives a complete set of n equations in n, which can be written in matrix form as

$$\left[\begin{array}{cc}[\mathbf{K}_{dd}] & [\mathbf{K}_{db}] \\ [\mathbf{K}_{bd}^{CC}] & [\mathbf{K}_{bb}^{CC}]\end{array}\right]\left\{\begin{array}{c}\{\mathbf{F}_d\} \\ \{\mathbf{F}_b\}\end{array}\right\}-\hat{\Omega}^2\left[\begin{array}{cc}[\mathbf{M}_{dd}] & [\mathbf{M}_{db}] \\ [\mathbf{M}_{bd}^{CC}] & [\mathbf{M}_{bb}^{CC}]\end{array}\right]\left\{\begin{array}{c}\{\mathbf{F}_d\} \\ \{\mathbf{F}_b\}\end{array}\right\}=[0] \tag{6.3.1}$$

Similarly, for S-S boundary condition, the governing equation is

$$\left[\begin{array}{cc}[\mathbf{K}_{dd}] & [\mathbf{K}_{db}] \\ [\mathbf{K}_{bd}^{SS}] & [\mathbf{K}_{bb}^{SS}]\end{array}\right]\left\{\begin{array}{c}\{\mathbf{F}_d\} \\ \{\mathbf{F}_b\}\end{array}\right\}-\hat{\Omega}^2\left[\begin{array}{cc}[\mathbf{M}_{dd}] & [\mathbf{M}_{db}] \\ [\mathbf{M}_{bd}^{SS}] & [\mathbf{M}_{bb}^{SS}]\end{array}\right]\left\{\begin{array}{c}\{\mathbf{F}_d\} \\ \{\mathbf{F}_b\}\end{array}\right\}=[0] \tag{6.3.2}$$

6.4 Results and Discussion

To analyse the impact of including surface effects along with geometric nonuniformity on the vibration behaviour of the considered nanobeam, the results for the non-dimensional frequency parameter $\Omega\left(=\omega\frac{L^2}{h}\sqrt{\frac{12\left(1-v^2\right)\rho_0^c}{E_0^c}}\right)$ have been obtained by solving equations (6.3.1) and (6.3.2) for C-C and S-S boundary conditions. In addition to the beam's parameter in Chapter 5, the values of the taper parameter is taken as: $\alpha = -0.5 \ldots 0.5$.

The comparison of the variation in the frequency parameter versus taper parameter without and with considering surface effects has been shown in Figures 6.2–6.4. In the absence of surface effect, the increasing values of the taper parameter, i.e., α, changes from –0.5 to 0.5, increase the frequency parameter for all values of gradient indices, slenderness ratio in all the modes, and for both the boundary conditions. This indicates that the increasing values of the taper parameter contribute to the stiffness of the tapered nanobeam, whatever the values of other parameters. The rate of increase is higher for the C-C beam as compared to the S-S beam. It increases further with the increase in the values of the slenderness ratio and modes of vibration.

However, in the presence of a surface effect, we observe both stiffening (hardening) and softening effects. This strongly depends upon the

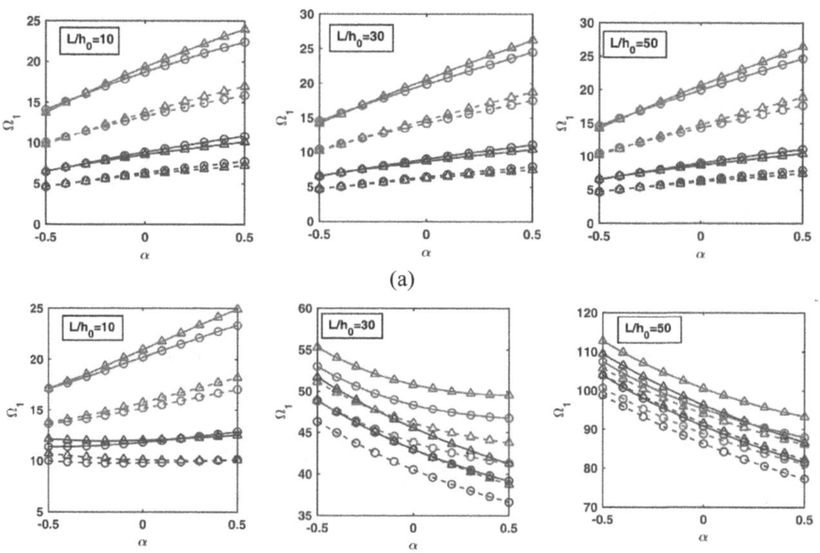

FIGURE 6.2
The variation of the frequency parameter versus nonuniformity parameter in the first mode of the BD-FG tapered nanobeams having $\mu = 2$: (a) WTS, and (b) WS; O, $p = 0$, Δ, $p = 2$; —, $n_z = 0$, — —, $n_z = 3$; Black, C-C, Gray, S-S.

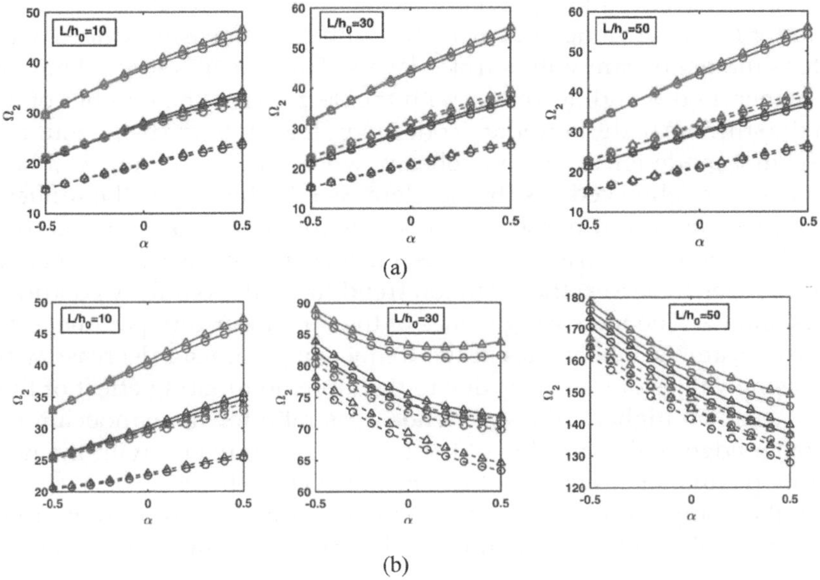

FIGURE 6.3
The variation of the frequency parameter versus nonuniformity parameter in the second mode of the BD-FG tapered nanobeams having $\mu = 2$: (a) WTS, and (b) WS; O, $p = 0$, Δ, $p = 2$; —, $n_z = 0$, — —, $n_z = 3$; Black, C-C, Gray, S-S.

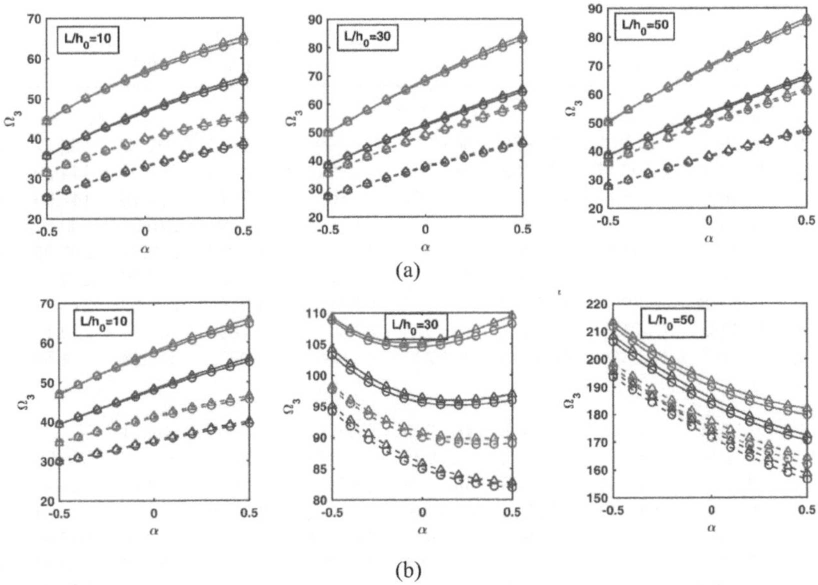

FIGURE 6.4
The variation of the frequency parameter versus nonuniformity parameter in the second mode of the BD-FG tapered nanobeams having $\mu = 2$: (a) WTS, and (b) WS; O, $p = 0$, Δ, $p = 2$; —, $n_z = 0$, — —, $n_z = 3$; Black, C-C, Gray, S-S.

slenderness ratio, boundary conditions and modes of vibration. For the lower values of L/h_0, the value of Ω increases with the increasing values of α and stiffens the nanobeam with surface layers. This effect is more dominant in the higher modes and decreases with the decrease in the number of modes and the stiffness of the boundary conditions. It leads to the behaviour change in the first mode for the S-S nanobeam, where both softening and hardening effects are observed. As the slenderness ratio increases, the influence of surface increases, and hardening and softening effects are noticed for both boundary conditions in all the modes. Further, an increase in the slenderness ratio changes (reverses) the vibration trend for both boundary conditions in all the modes. The increasing values of the nonuniformity parameter soften the nanobeam, i.e., the values of the frequency parameter decrease with the increasing values of α. This is due to the more dominating effect of the surface energy on higher values of slenderness ratio for the nanobeam thin at the outer edge. This variation of the frequency parameter with the nonuniformity parameter is significant for the BD-FG nanobeam.

Another comparative study on the impact of surface effect on vibration behaviours of the BD-FG tapered nanobeam has been elaborated in Tables 6.1–6.3,

TABLE 6.1

The Values of Ω for BD-FG Tapered Nanobeam with Surface Layers Having $\mu = 2$, $L/h_0 = 30$ in the First Mode

		WTS			WS			Diff %		
						α				
n_z	n_x	−0.3	0	0.3	−0.3	0	0.3	−0.3	0	0.3
						C-C				
0	0	16.8041	19.8714	22.7030	50.6361	48.3407	47.1173	201.33	143.27	107.54
2	1	16.6966	20.0395	23.1483	51.1532	48.9546	47.8494	206.37	144.29	106.71
5	2	16.8697	20.5597	24.0227	52.9884	50.8003	49.7614	214.10	147.09	107.14
	0	12.4462	14.7157	16.8097	46.8239	44.2228	42.5396	276.21	200.51	153.07
	1	12.3666	14.8402	17.1393	47.3492	44.8116	43.1982	282.88	201.96	152.04
	2	12.4947	15.2252	17.7866	49.1407	46.5783	44.9723	293.29	205.93	152.84
	0	11.5018	13.5964	15.5278	45.9650	43.3119	41.5413	299.63	218.55	167.53
	1	11.4282	13.7113	15.8321	46.4915	43.8957	42.1857	306.81	220.14	166.46
	2	11.5465	14.0669	16.4298	48.2732	45.6467	43.9331	318.08	224.50	167.40
						S-S				
0	0	7.5774	9.0013	10.2875	46.1998	42.9682	40.4995	509.71	377.36	293.68
1	1	7.6046	8.9245	10.1082	46.8925	43.6144	41.0463	516.63	388.70	306.07
5	2	7.5025	8.6963	9.7585	48.9519	45.5454	42.8037	552.47	423.73	338.63
	0	5.6141	6.6689	7.6215	43.9649	40.7632	38.2430	683.12	511.24	401.78
	1	5.6343	6.6120	7.4886	44.5935	41.3801	38.8058	691.46	525.83	418.20
	2	5.5586	6.4428	7.2294	46.5265	43.2228	40.5258	737.02	570.87	460.57
	0	5.1902	6.1650	7.0453	43.4626	40.2734	37.7486	737.40	553.26	435.80
	1	5.2088	6.1124	6.9225	44.0775	40.8832	38.3136	746.21	568.86	453.46
	2	5.1388	5.9559	6.6827	45.9822	42.7047	40.0224	794.80	617.02	498.90

TABLE 6.2

The Values of Ω for BD-FG Tapered Nanobeam with Surface Layers Having $\mu = 2$, $L/h_0 = 30$ in the Second Mode

		WTS			WS			Diff %		
					α					
n_z	n_x	−0.3	0	0.3	−0.3	0	0.3	−0.3	0	0.3
					C-C					
0	0	36.9315	43.6081	49.6426	84.4216	81.7460	81.2141	128.59	87.46	63.60
2	1	36.8101	43.7731	50.0894	84.6167	82.1160	81.7627	129.87	87.59	63.23
5	2	36.9662	44.2699	50.9177	85.6227	83.2337	83.0236	131.62	88.01	63.05
	0	27.3424	32.2749	36.7286	77.1650	73.3912	71.4598	182.22	127.39	94.56
	1	27.2525	32.3971	37.0592	77.3857	73.7486	71.9395	183.96	127.64	94.12
	2	27.3681	32.7647	37.6720	78.3991	74.8359	73.1013	186.46	128.40	94.05
	0	25.2544	29.7983	33.8960	75.5914	71.6260	69.4166	199.32	140.37	104.79
	1	25.1714	29.9111	34.2010	75.8170	71.9817	69.8846	201.20	140.65	104.33
	2	25.2781	30.2505	34.7665	76.8319	73.0658	71.0322	203.95	141.54	104.31
					S-S					
0	0	24.7628	29.2758	33.4535	78.1820	73.8917	71.5387	215.72	152.40	113.85
1	1	24.7241	29.3819	33.7045	78.5730	74.2583	71.8650	217.80	152.73	113.22
5	2	24.8641	29.7103	34.2230	79.7766	75.3673	72.8830	220.85	153.67	112.96
	0	18.3423	21.6821	24.7727	73.1872	68.4832	65.3473	299.01	215.85	163.79
	1	18.3136	21.7607	24.9584	73.5304	68.8418	65.6746	301.51	216.36	163.14
	2	18.4172	22.0038	25.3422	74.6663	69.9323	66.6765	305.42	217.82	163.10
	0	16.9518	20.0351	22.8868	72.1110	67.3447	64.0582	325.39	236.13	179.89
	1	16.9253	20.1077	23.0583	72.4421	67.7017	64.3897	328.01	236.70	179.25
	2	17.0210	20.3321	23.4126	73.5615	68.7889	65.3939	332.18	238.33	179.31

TABLE 6.3

The Values of Ω for BD-FG Tapered Nanobeam with Surface Layers Having $\mu = 2$, $L/h_0 = 30$ in the Third Mode

		WTS			WS			Diff %		
					α					
n_z	n_x	−0.3	0	0.3	−0.3	0	0.3	−0.3	0	0.3
					C-C					
0	0	57.7364	67.9953	77.0808	105.5588	104.5222	106.2306	82.83	53.72	37.82
2	1	57.6321	68.1283	77.4447	105.6224	104.7171	106.5765	83.27	53.71	37.62
5	2	57.7529	68.5284	78.1125	106.1189	105.2893	107.2917	83.75	53.64	37.36
	0	42.7225	50.2875	56.9765	94.6007	91.2170	90.3562	121.43	81.39	58.59
	1	42.6454	50.3859	57.2456	94.7005	91.4172	90.6464	122.06	81.43	58.35
	2	42.7349	50.6819	57.7394	95.2783	92.0064	91.2782	122.95	81.54	58.09
	0	39.4339	46.3867	52.5223	92.2967	88.4765	87.0734	134.05	90.74	65.78
	1	39.3628	46.4775	52.7704	92.4048	88.6833	87.3592	134.75	90.81	65.55
	2	39.4455	46.7507	53.2259	93.0095	89.2945	87.9912	135.79	91.00	65.32

(Continued)

TABLE 6.3 (Continued)

The Values of Ω for BD-FG Tapered Nanobeam with Surface Layers Having $\mu = 2$, $L/h_0 = 30$ in the Third Mode

		WTS			WS			Diff %		
					α					
n_z	n_x	−0.3	0	0.3	−0.3	0	0.3	−0.3	0	0.3
					S-S					
0	0	44.3334	52.3684	59.6074	98.8708	95.5916	95.2321	123.02	82.54	59.77
1	1	44.2941	52.4764	59.8598	99.0924	95.7974	95.4671	123.71	82.55	59.48
5	2	44.4370	52.8022	60.3616	99.7856	96.4066	96.0869	124.56	82.58	59.19
	0	32.8250	38.7628	44.1075	90.5050	85.8668	83.6678	175.72	121.52	89.69
	1	32.7959	38.8427	44.2942	90.7217	86.0813	83.8747	176.63	121.62	89.36
	2	32.9017	39.0837	44.6653	91.4428	86.7202	84.4585	177.93	121.88	89.09
	0	30.3212	35.7930	40.7127	88.7393	83.8727	81.2924	192.66	134.33	99.67
	1	30.2943	35.8667	40.8850	88.9519	84.0926	81.5001	193.63	134.46	99.34
	2	30.3920	36.0891	41.2272	89.6807	84.7499	82.0902	195.08	134.84	99.12

by reporting the values of the frequency parameter and percentage increase in the frequency parameter due to consideration of surface effect for different values of taper parameter, functionally gradient index, heterogeneity index in all three modes, and for both boundary conditions. For a uniform beam, the results show that the consideration of surface effects stiffens the nanobeam, hence increasing frequencies. As the beam is graded along the thickness direction, i.e. n_z increases from 0 to 5, the influence of the surface effect increases significantly in the first mode. This effect further increases as the material properties are also graded in the axial direction.

Now, as the nonuniformity in the nanobeam arises and the beam changes from thinner to thicker towards the outer, the value of the frequency parameter increases in the absence of the surface effect. However, in the presence of surface effect, reverse behaviour has been noticed. The percentage increase in the values of Ω due to the change in the values of the taper parameter from $\alpha = -0.3$ to $\alpha = 0.3$ for $\mu = 2$, $n_z = 5$, $p = 1$, $L/h_0 = 30$ is 42.29, −8.99 for the C-C and 37.54, −12.96 for S-S nanobeam in the absence and presence of surface effect, respectively. Thus, the nanobeam with surface layers has a completely reverse trend of stiffness variation with nonuniformity parameter as compared to the nanobeam without a surface layer. This is attributed to the dominating effect of the surface on the beam, thinner at the outer edge, than the beam is thicker at the outer edge. The effect of surface energy is more pronounced for the S-S beam as compared to the C-C beam. This effect is higher in the first mode and significantly decreases with the increasing modes.

6.5 Summary

In summary, this chapter demonstrates the effective application of nonlocal elasticity and surface elasticity theories in comprehensively characterizing the behaviour of the non-uniform BD-FG nanobeams. The analysis highlights the critical role played by the interplay of nonlocal effects, surface stresses, material gradation, and geometric nonuniformity in the accurate design and control of smart nanostructures. The findings of this study provide valuable insights for the design and optimization of advanced nanodevices incorporating nanostructure-based sensors.

References

Dangi, Chinika, Roshan Lal, and N. Sukavanam. 2021. "Effect of Surface Stresses on the Dynamic Behaviour of Bi-Directional Functionally Graded Nonlocal Strain Gradient Nanobeams via Generalized Differential Quadrature Rule." *European Journal of Mechanics, A/Solids* 90(March):104376. doi: 10.1016/j.euromechsol.2021.104376.

Despenić, Nikola, Goran Janevski, and Ivan Pavlović. 2023. "Vibrations of Nonuniform Bidirectional Functionally Graded Nanotubes Based on the Refined Beam Theory in a Thermal Environment." *Journal of Mechanics of Materials and Structures* 18(1):59–74. doi: 10.2140/jomms.2023.18.59.

Karami, Behrouz, Maziar Janghorban, and Timon Rabczuk. 2020. "Dynamics of Two-Dimensional Functionally Graded Tapered Timoshenko Nanobeam in Thermal Environment Using Nonlocal Strain Gradient Theory." *Composites Part B: Engineering* 182(October 2019):107622. doi: 10.1016/j.compositesb.2019.107622.

Lal, Roshan, and Chinika Dangi. 2019. "Thermomechanical Vibration of Bi-Directional Functionally Graded Non-Uniform Timoshenko Nanobeam Using Nonlocal Elasticity Theory." *Composites Part B: Engineering* 172(May):724–42. doi: 10.1016/j.compositesb.2019.05.076.

Lal, Roshan, and Chinika Dangi. 2021a. "Dynamic Analysis of Bi-Directional Functionally Graded Timoshenko Nanobeam on the Basis of Eringen's Nonlocal Theory Incorporating the Surface Effect." *Applied Mathematics and Computation* 395:125857. doi: 10.1016/j.amc.2020.125857.

Lal, Roshan, and Chinika Dangi. 2021b. "Effect of In-Plane Load and Thermal Environment on Buckling and Vibration Behaviour of Two-Dimensional Functionally Graded Tapered Timoshenko Nanobeam." *Journal of Vibration and Acoustics* 143(1):1–14. doi: 10.1115/1.4047862.

Mukhopadhyay, Tanmoy, Avik Mahata, Susmita Naskar, and Sondipon Adhikari. 2020. "Probing the Effective Young's Modulus of 'Magic Angle' Inspired

Multi-Functional Twisted Nano-Heterostructures." *Advanced Theory and Simulations* 3(10):1–17. doi: 10.1002/adts.202000129.

Saumya, Kumar, Susmita Naskar, and Tanmoy Mukhopadhyay. 2023. "'Magic' of Twisted Multi-Layered Graphene and 2D Nano-Heterostructures." *Nano Futures* 7(3). doi: 10.1088/2399-1984/acf0a9.

Shafiei, Navvab, and Mohammad Kazemi. 2017. "Buckling Analysis on the Bi-Dimensional Functionally Graded Porous Tapered Nano-/Micro-Scale Beams." *Aerospace Science and Technology* 66:1–11. doi: 10.1016/j.ast.2017.02.019.

7

Nonlocal Surface Model of Vibrations and Buckling of BD-FG Nanobeam Embedded with Piezoelectric Layers

7.1 Overview

Functionally graded piezoelectric (FGP) nanostructures have gained significant attention as a key design element for enhancing the performance of devices operating under complex mechanical, electrical, and thermal loading conditions (El-Galy, Saleh, and Ahmed 2019; Singh et al. 2023). The integration of piezoelectric nanostructures with NEMs/MEMs, sensors, and actuators has enabled innovative applications across various fields of science and engineering (Malakooti, Zhou, and Sodano 2018; Siang, Lim, and Salman Leong 2018; Wang et al. 2023). Recent studies have explored the behaviour of FGP nanostructures (Li et al. 2021; Marzbanrad et al. 2018; Yu et al. 2021), emphasizing the crucial role of material inhomogeneity in determining their electro-mechanical characteristics. Existing literature shows that multidirectional FGMs can endure large deformations, strong electric fields, and wide temperature variations (Derayatifar, Tahani, and Moeenfard 2017), which has motivated extensive research into the static and dynamic behaviour of BD-FG nanostructures. This has led to the numerous studies on the static and dynamic analysis of bi-directional functionally graded (BD-FG) nanostructures. Recent work in this area can be found in references (Rahmani, Faroughi, and Friswell 2020; Sahmani and Safaei 2019; Yang et al. 2018). Moreover, very selective studies are available on BD-FG nanostructures considering the influence of surface effects. Attia and Shanab (2022) investigated the dynamic response of BD-FG porous nanobeams subjected to a moving harmonic load, incorporating both surface elasticity theory (SET) and modified couple stress theory (MCST). The vibration behaviour of BD-FG nanobeams considering surface stress was studied by C. Dangi et al. using the ENT [Lal, and Dangi 2021] and NSGT (Dangi, Lal, and Sukavanam 2021). Additionally, Dangi and Naskar (2025) analysed the combined effects of non-uniform geometry, surface stress, and nonlocal elasticity on the vibration characteristics of BD-FG tapered nanobeams. Although several studies have

DOI: 10.1201/9781003583806-7

been conducted, the comprehensive studies addressing the combined effects of surface energy and nonlocal elasticity on responses of FGP nanostructures are very limited. Recently, the nonlinear vibration of the FGP nanobeams has been studied by Arefi et al. (2019), based on the nonlocal elasticity theory considering surface and flexoelectric effects, and reported that considering flexoelectricity. A work on vibration and bending behaviour of FGP considering flexoelectric and surface effect has been presented by Wang et al. (2025).

This chapter aims to investigate the underlying mechanisms of the coupled influence of surface energy and nonlocal effects on the electromechanical response of BD-FGP nanostructures subjected to an applied voltage. A BD-FGP nanobeam with surface layers is considered, and an external voltage is applied across its thickness. To establish an accurate mathematical framework, electro-mechanical stresses are modelled in conjunction with surface and nonlocal stress effects. Hamilton's principle is used to derive the governing equations and associated boundary conditions. These nonlocal equations of motion are then discretized and solved using the GDQM to obtain the natural frequencies. A detailed parametric study follows, investigating how variations in beam parameters and external voltage influence the vibration behaviour of the BD-FGP nanobeam.

7.2 Mathematical Formulation of Vibrations for BD-FGP Nanobeam

A BD-FGP nanobeam with surface layers subjected to external electric voltage has been considered for investigation, as shown in Figure 7.1. The nanobeam is assumed to be composed of piezoelectric material PZT-5 and Aluminium Al as ceramic and metallic constituents of the beam, respectively (Derayatifar et al. 2017). The material properties of the beam are graded in the thickness direction according to a power law function and as per the exponential function in the length direction (Birman and Byrd 2007; Parihar, Setti, and Sahu

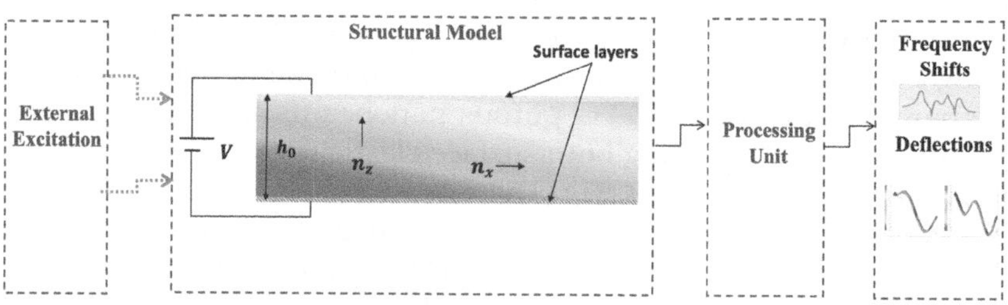

FIGURE 7.1
Schematic view of BD-FGP nanobeam with surface layers subjected to an external field.

2018). The upper surface $\left(S^{+}\right)$ of the beam at $z = h/2$ is ceramic-rich, while the bottom surface $\left(S^{-}\right)$ at $z = -h/2$ is metallic rich. An external electric voltage V has been applied across the thickness of the beam. The specific material property $P(x, y, z)$ of the BD-FG piezoelectric nanobeam at any point is given by (Karami, Janghorban, and Rabczuk 2020),

$$P(x,y,z) = e^{n_x \frac{x}{L}} \left[(P_c - P_m) \left(\frac{z}{h} + \frac{1}{2} \right)^{n_z} + P_m \right] \tag{7.2.1}$$

7.2.1 Nonlocal Theory of Piezoelectricity with Surface Effect

As per the Euler-Bernoulli theory (Karami, Janghorban, and Rabczuk 2020), the displacement components (u, v, w) of an arbitrary material point $p(x, y, z)$ along the axes are given by

$$u(x,z,t) = -z \frac{\partial w_0}{\partial x}, \quad v = 0 \quad w(x,z,t) = w_0(x,t) \tag{7.2.2}$$

where w_0 represent the transverse displacement of the mid-plane.

The linear constitutive relationship defined for electrical and mechanical interactions in piezo-electric materials is (Penna and Lovisi 2025),

$$\left. \begin{aligned} \sigma &= C : \varepsilon - e : E \\ D &= e^T : \varepsilon - \eta : E \end{aligned} \right\} \tag{7.2.3}$$

where, C is the elastic constant matrix, e is the piezoelectric constants, η is the dielectric constants. Tensors σ, ε, E, D are the classical stress, strain, electric field, and electric displacement, respectively.

The electric field components are

$$E = \begin{bmatrix} E_x \\ E_y \\ E_z \end{bmatrix} = \begin{bmatrix} -\dfrac{\partial \phi}{\partial x} \\ -\dfrac{\partial \phi}{\partial y} \\ -\dfrac{\partial \phi}{\partial z} \end{bmatrix} \tag{7.2.4}$$

where, the electric potential across the thickness is $\phi = \frac{2V}{h} Z$. The potential is $\phi = V$ at the top and $\phi = 0$ at the bottom surface. Since ϕ is a function of z only, then $E_x \lll E_z$. This leads to the vanishing axial dielectric displacement, i.e., $D_x = 0$.

Therefore, the constitutive relation for a thin piezoelectric nanobeam in equation (7.2.3) gives

$$\sigma_{xx} = \frac{C_{11}(x,z)}{1-\nu^2}\varepsilon_{xx} - e_{13}(x,z)E_z, \quad D_z = e_{13}(x,z)\varepsilon_{xx} - \eta_{33}(x,z)E_z \quad (7.2.5)$$

7.2.2 Surface Elasticity Theory

Surface elasticity theory proposed by Gurtin-Murdoch is the most widely used and efficient continuum theory to incorporate the effect of surface energy in the responses of nanostructures. According to the GM-SET (Attia 2017), a solid is assumed to be composed of a bulk part and surface layers. The surface layers are simulated as membranes of negligible thickness with different material properties from the underlying bulk and perfectly bonded to the bulk part. The constitutive electro-mechanical stress-strain relations for the two surface layers are

$$\sigma_{xx}^{S\pm} = \sigma_0^{S\pm} + C_{11}^{S\pm}(x)\varepsilon_{xx}^{S\pm} - e_{13}^{S\pm}(x)E_z^{S\pm}, \quad t_{xz}^{s\pm} = \tau_0^{s\pm}\frac{\partial w_0(x,t)}{\partial x} \quad (7.2.6)$$

where, $\sigma_0^{S\pm}$ is the residual surface stress, $C_{11}^{S\pm}$ is the surface elasticity, $\tau_0^{s\pm}$ are residual surface tension, $e_{13}^{S\pm}$ is the surface piezoelectric constant, and $\varepsilon_{xx}^{s\pm}$ are strain components of the surface $S^{\pm}$.

Using G-M surface elasticity, the modified transverse stress is given as

$$\begin{aligned}\sigma_{zz} &= \frac{1}{2}\left[\left(\tau_0^{s+} - \tau_0^{s-}\right)\frac{\partial^2 w_0}{\partial x^2} - \left(\rho^{s+} - \rho^{s-}\right)\frac{\partial^2 w_0}{\partial t^2}\right] \\ &+ \frac{z}{h}\left[\left(\tau_0^{s+} + \tau_0^{s-}\right)\frac{\partial^2 w_0}{\partial x^2} - \left(\rho^{s+} + \rho^{s-}\right)\frac{\partial^2 w_0}{\partial t^2}\right]\end{aligned} \quad (7.2.7)$$

Therefore, the classical constitutive relation for the bulk is

$$\begin{aligned}\sigma_{xx} &= \frac{C_{11}(x,z)}{1-\nu^2}\varepsilon_{xx} - e_{13}(x,z)E_z + \frac{\nu}{(1-\nu)}\sigma_{zz}, \\ D_z &= e_{13}(x,z)\varepsilon_{xx} - \eta_{33}(x,z)E_z\end{aligned} \quad (7.2.8)$$

7.2.3 Eringen's Nonlocal Elasticity Theory

According to ENT (Eringen 1983), the stress at any point in the body depends not only on the strain at that point but also on the strain at all other neighbouring points. The differential form of the nonlocal constitutive electro-mechanical relation is

$$\left(1-\mu\nabla^2\right)\sigma_{xx}^{nl} = \frac{C_{11}(x,z)}{1-\nu^2}\varepsilon_{xx} - e_{13}(x,z)E_z + \frac{\nu}{(1-\nu)}\sigma_{zz},$$
$$\left(1-\mu\nabla^2\right)D_z^{nl} = e_{13}(x,z)\varepsilon_{xx} - \eta_{33}(x,z)E_z \tag{7.2.9}$$

where $\mu = (e_0\bar{a})^2$ is known as the nonlocal parameter; vectors $\boldsymbol{\sigma}^{nl}, \boldsymbol{D}^{nl}$ are the nonlocal stress and dielectric displacement vectors, respectively.

Similarly, the nonlocal stresses for surface layers are

$$\left(1-\mu\nabla^2\right)\sigma_{xx}^{S\pm} = \sigma_0^{S\pm} + C_{11}^{S\pm}(x)\varepsilon_{xx}^{S\pm} - e_{13}^{S\pm}(x)E_z^{S\pm}, \quad \left(1-\mu\nabla^2\right)t_{xz}^{s\pm} = \tau_0^{s\pm}\frac{\partial w_0(x,t)}{\partial x} \tag{7.2.10}$$

Following Hamilton's Energy Principle, the Euler-Lagrange equation for the considered beam is obtained as follows:

$$\frac{\partial^2 M}{\partial x^2} + \frac{\partial Q_x^S}{\partial x} + \frac{\partial}{\partial x}\left(N\frac{\partial w_0}{\partial x}\right) = I_0\frac{\partial^2 w_0}{\partial t^2} - \frac{\partial}{\partial x}\left(I_2\frac{\partial^2 w_0}{\partial t^2}\right) \tag{7.2.11}$$

where M, Q_x^S, N, I_0, and I_2 are nonlocal moments, axial loads, and mass moments of inertia, respectively. The nonlocal moments for a nanobeam with a surface layer are given by

$$M = \int_{-h/2}^{h/2} b\ z\ \sigma_{xx}^{nl}dz - \int_{-h/2}^{h/2} b\ z\ D_z^{nl}dz + zb\sigma_{xx}^{S+}\Big|_{\pm\frac{h}{2}},\ Q_x^S$$
$$= \int_{-h/2}^{h/2} b\left(t_{xz}^{s\pm} + t_{xz}^{s\pm}\right)dz,\ N = 2be_{13}V$$

$$M = \int_{-h/2}^{h/2} bz\sigma_{xx}^{nl}dz - \int_{-h/2}^{h/2} bzD_z^{nl}dz + zb\sigma_{xx}^{S+}\Big|_{\pm\frac{h}{2}},$$

$$Q_x^S = \int_{-h/2}^{h/2} b\left(t_{xz}^{s\pm} + t_{xz}^{s\pm}\right)dz, \quad N = 2be_{13}V \tag{7.2.12}$$

$$\left(I_0(x), I_2(x)\right) = \int_{-h/2}^{h/2} b\rho(x,z)\left(1,z^2\right)dz + \left(b\left(\rho^{S+}+\rho^{S-}\right), \frac{h^2 b\left(\rho^{S+}+\rho^{S-}\right)}{4}\right)$$

Now, the nonlocal moments in terms of deflection are

$$\left(1-\mu^2\frac{\partial^2}{\partial x^2}\right)M = A_1\frac{\partial^2 w_0}{\partial x^2} - S_1\frac{\partial^2 w_0}{\partial t^2} + C_1 \tag{7.2.13}$$

$$\left(1-\mu^2\frac{\partial^2}{\partial x^2}\right)Q_x^S = \left(\tau_0^{s+}+\tau_0^{s-}\right)\frac{\partial w_0}{\partial x} \tag{7.2.14}$$

where $A_1 = \frac{b}{(1-v^2)}\int_{-h/2}^{h/2}\left(-z^2C_{11}\right)dz - b\int_{-h/2}^{h/2} z^2\left(\frac{e_{13}}{\eta_{33}}\right)dz + b\int_{-h/2}^{h/2} ze_{13}\frac{V}{h}dz$

$$-\frac{h^2b(C_{11}^{s+}+C_{11}^{s-})}{4}+\frac{h^2be_{13}(e_{13}^{s+}+e_{13}^{s-})}{4\eta_{33}}+b\frac{(e_{13}^{s+}+e_{13}^{s-})}{2}V+b\int_{-h/2}^{h/2}\frac{z^2}{h}\frac{v}{(1-v)}\left(\tau_0^{s+}+\tau_0^{s-}\right)dz;$$

$$S_1 = b\int_{-h/2}^{h/2}\frac{z^2}{h}\frac{v}{(1-v)}\left(\rho_0^{s+}+\rho_0^{s-}\right)dz;\ C_1 = \frac{h}{2}\left(\tau_0^{s+}-\tau_0^{s-}\right).$$

The substitution of nonlocal moments from the above equations (7.2.12)–(7.2.14) into equation (7.2.11) gives rise to a nonlocal governing equation of motion for free vibration of BD-FGP nanobeam with surface layers as follows:

$$A_1\frac{\partial^4 w_0}{\partial x^4}+2\frac{\partial A_1}{\partial x}\frac{\partial^3 w_0}{\partial x^3}+\frac{\partial^2 A_1}{\partial x^2}\frac{\partial^2 w_0}{\partial x^2}+\left(1-\mu\frac{\partial^2}{\partial x^2}\right)\frac{\partial}{\partial x}\left(N\frac{\partial w_0}{\partial x}\right)+\left(\tau_0^{s+}+\tau_0^{s-}\right)\frac{\partial^2 w_0}{\partial x^2}$$

$$=\left(1-\mu\frac{\partial^2}{\partial x^2}\right)\left(I_0(x)\frac{\partial^2 w_0}{\partial t^2}-\frac{\partial}{\partial x}\left(I_2(x)\frac{\partial^3 w_0}{\partial x\partial t^2}\right)\right)+S_1\frac{\partial^4 w_0}{\partial x^2\partial t^2} \tag{7.2.15}$$

For harmonic vibrations, we have $w_0 = e^{i\omega t}\overline{w}_0$

$$A_1\frac{d^4\overline{w}_0}{dx^4}+2\frac{dA_1}{dx}\frac{d^3\overline{w}_0}{dx^3}+\frac{d^2A_1}{dx^2}\frac{d^2\overline{w}_0}{dx^2}+\frac{dN}{dx}\frac{d\overline{w}_0}{dx}+N\frac{d^2\overline{w}_0}{dx^2}$$

$$-\mu\left(N\frac{d^4\overline{w}_0}{dx^4}+3\frac{dN}{dx}\frac{d^3\overline{w}_0}{dx^3}+3\frac{d^2N}{dx^2}\frac{d^2\overline{w}_0}{dx^2}+\frac{d^3N}{dx^3}\frac{d\overline{w}_0}{dx}\right)+\left(\tau_0^{s+}+\tau_0^{s-}\right)\frac{d^2\overline{w}_0}{dx^2}$$

$$=-\omega^2\left[I_0\overline{w}_0-\frac{d}{dx}\left(I_2\frac{d\overline{w}_0}{dx}\right)-\mu\left(I_0\frac{d^2\overline{w}_0}{dx^2}+2\frac{dI_0}{dx}\frac{d\overline{w}_0}{dx}+\frac{d^2I_0}{dx^2}\overline{w}_0\right)\right.$$

$$\left.+\mu\left(I_2\frac{d^4\overline{w}_0}{dx^4}+3\frac{dI_2}{dx}\frac{d^3\overline{w}_0}{dx^3}+3\frac{d^2I_2}{dx^2}\frac{d^2\overline{w}_0}{dx^2}+\frac{d^3I_2}{dx^3}\frac{d\overline{w}_0}{dx}\right)+S_1\frac{d^2\overline{w}_0}{dx^2}\right] \tag{7.2.16}$$

The above equation has been solved numerically for the first three modes of vibration with different boundary conditions.

Using GDQM, the discretized form of the governing equation, together with boundary conditions, leads to a complete eigenvalue problem as:

$$\begin{bmatrix} [K_{dd}] & [K_{db}] \\ [K_{bd}] & [K_{bb}] \end{bmatrix} \begin{Bmatrix} \{W_d\} \\ \{W_b\} \end{Bmatrix} - \omega^2 \begin{bmatrix} [M_{dd}] & [M_{db}] \\ [M_{bd}] & [M_{bb}] \end{bmatrix} \begin{Bmatrix} \{W_d\} \\ \{W_b\} \end{Bmatrix} = [0] \tag{7.2.17}$$

7.3 Results and Discussion

In this section, the combined impact of piezoelectricity and surface parameters on the vibration characteristics of the BD-FG nanobeam has been examined. In order to accomplish this, the results for the non-dimensional frequency parameter $\Omega\left(=\omega\frac{L^2}{h}\sqrt{\frac{12(1-v^2)\rho_0^c}{E_0^c}}\right)$ have been simulated by developing a computer program for various values of material indices, applied voltage, and nonlocal parameter, subjected to different types of boundary conditions. The constituent materials for the considered beam are PZT-5 and Al, and their material properties are given in Table 7.1. In addition to the values of beam parameters specified in Chapter 5, the values of Electric Voltage are: $V = -0.5\ldots0.5$

For a comprehensive investigation into the vibration characteristics of the nanobeam, the influence of key parameters affecting its stability under electrical loading is illustrated and discussed in Figures 7.2 and 7.3.

The influence of nonlocal and surface parameters on the vibration responses of the BD-FGP nanobeam subjected to electric voltage has been discussed in Figure 7.2. The graphs of the frequency parameter versus the nonlocal parameter have been plotted, in the absence as well as the presence of external voltage, for C-C, C-S, and S-S nanobeams. The results indicate a trend of decreasing frequency with increasing nonlocal parameter for all boundary conditions, with the exception of the first vibration mode of the S-S nanobeam under an applied voltage of $V = 0.2$. This is attributed to the

TABLE 7.1

Material Properties of Aluminium (Al) and PZT-5

Material	E	ρ	ρ_0	τ_0	E_s	e_{31}	η_{33}
PZT-5	126	7,500	3.17e-7	1	7.56	−6.5	1.3e-7
A1	70	2,700	5.46e-7	0.9108	5.1882	−10	1.0275e-8

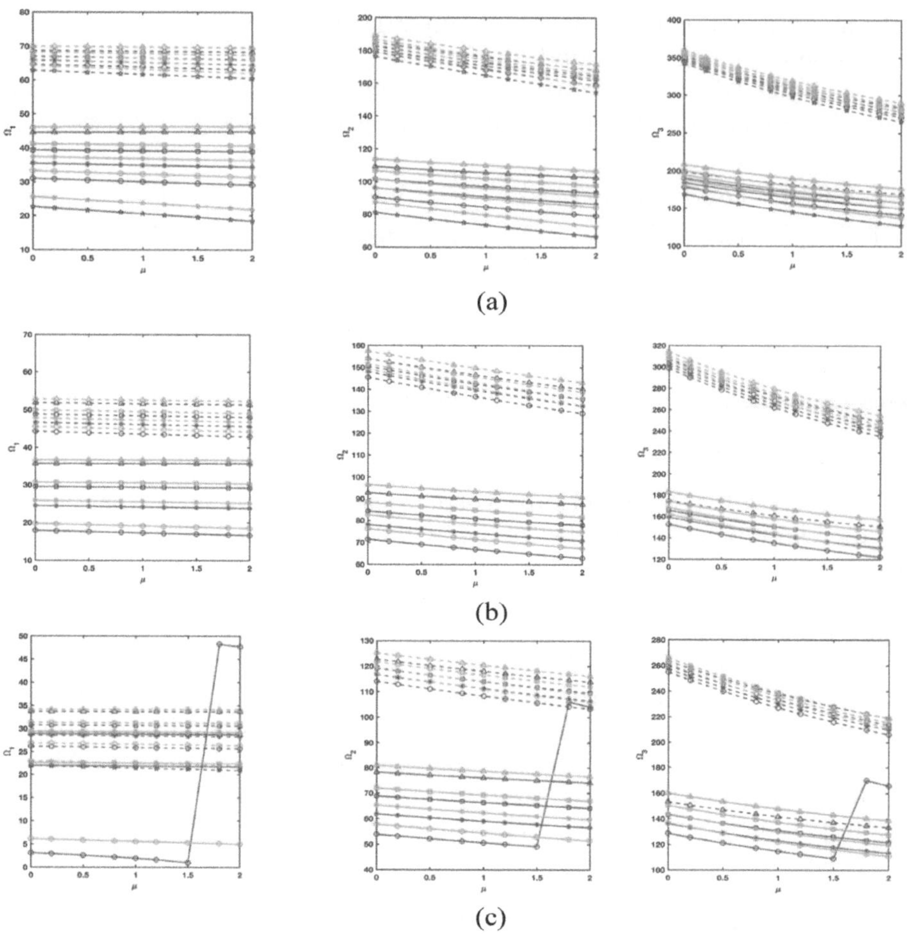

FIGURE 7.2
The values of Ω versus μ for BD-FGP nanobeams having $n_z = 1$: Δ, $V = -0.5$; $\square$, $V = -0.2$; $*$, $V = 0$; O, $V = 0.2$; *Black*, $n_x = 0$; *Gray*, $n_x = 2$; –, WTSE; —, WSE; (a) C-C, (b) C-S, (c) S-S.

reduction in beam stiffness, which significantly lowers the natural frequency and may ultimately lead to structural failure due to large amplitude of the vibrations.

Incorporating the bi-directional gradation in material properties enhances the sensitivity of the nanobeam to nonlocal effects, particularly under positive external voltage. This influence becomes even more significant in the presence of surface effects. Among the three boundary conditions, the nonlocal parameter shows the most pronounced effect in the C-C beam, which is dominant in higher modes of vibration.

The impact of electric voltage V on the frequency response of the BD-FGP nanobeam is illustrated in Figure 7.3. It represents the variation of the

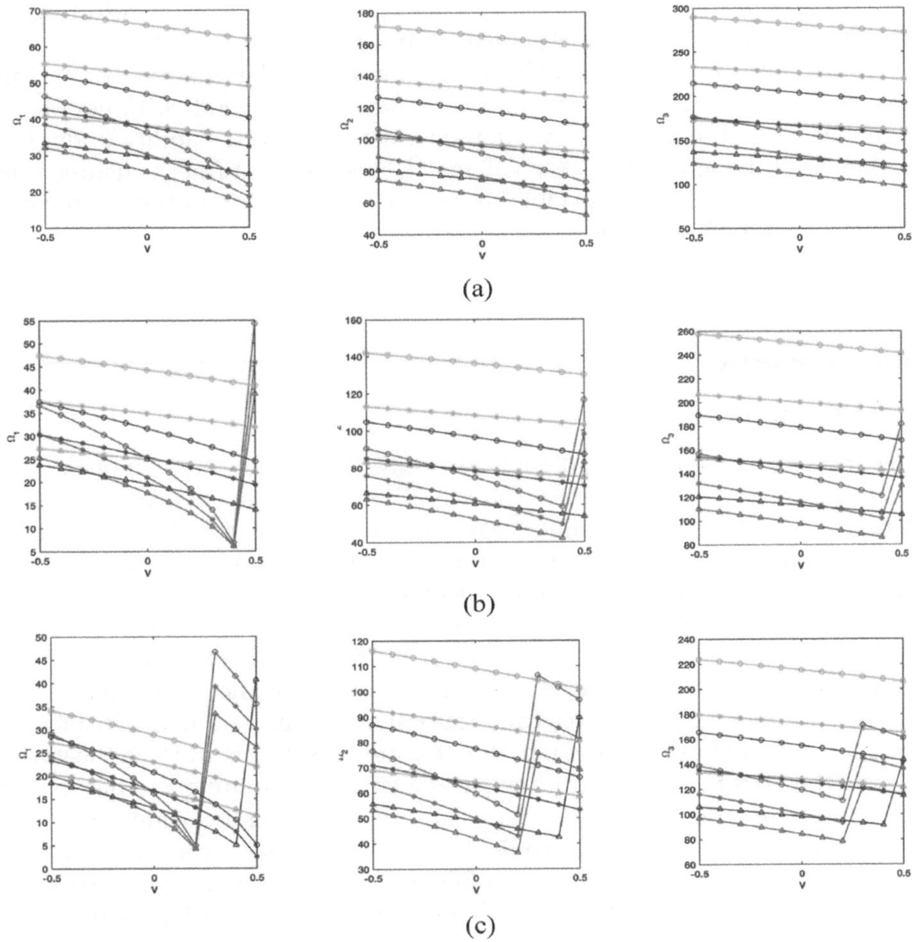

FIGURE 7.3
The values of Ω versus V for BD-FGP nanobeams having $\mu = 1$: Δ, $n_z = 0$; *, $n_z = 1$; O, $n_z = 5$; *Black*, $n_x = 0$; *Dark Gray*, $n_x = 1$; *Light Gray*, $n_x = 2$; (a) C-C, (b) C-S, (c) S-S.

frequency parameter with the applied voltage for all three boundary conditions, both with and without bi-directional material gradation.

The results show that the frequency parameter decreases as the applied voltage varies from $V = -0.5$ to $V = 0.5$. This reduction is primarily attributed to the decrease in the effective stiffness of the beam under positive electric voltage. The effect becomes more pronounced when the beam is graded along the thickness direction, particularly for a high transverse gradient index $(n_z = 5)$. Among the three boundary conditions, the S-S beam exhibits the steepest decline in the frequency parameter compared to C-S and C-C beams. This reduction is especially critical for a homogeneous nanobeam

($n_x = 0$), where the frequency approaches zero as near $V = 0.5$, resulting in structural instability for the S-S and C-S configurations.

However, the introduction of bi-directional material gradation by varying the axial gradient index n_x from 0 to 2 enhances the overall stiffness of the nanobeam. As a result, the beam demonstrates improved stability under the influence of the applied electric voltage V. This added stiffness ensures that the nanobeam maintains a stable frequency response across the entire voltage range from $V - 0.5$ to $V = 0.5$.

7.4 Summary

In summary, this chapter systematically outlines the development of a mathematical framework to analyse the vibration. This chapter presents a systematic derivation of the nonlocal governing differential equations of motion for BD-FGP nanobeams, enabling a more generalized and versatile framework for material design. The analysis highlights the significant influence of applied electric fields on the natural frequencies of the system. The results reveal complex vibrational behaviour arising from the electro-mechanical coupling in BD-FGP nanobeams. Notably, under specific parameter conditions, the applied electric force can induce dynamic or parametric instability by substantially lowering the vibration frequencies. Depending on the variation in the axial functional grading parameter, both softening and hardening responses are observed. However, by carefully selecting the beam configuration and material gradation profile, these adverse effects can be mitigated. Finally, the study demonstrates that BD-FGP nanostructures can be strategically engineered to optimize their dynamic response to electric fields, offering promising implications for the design of next-generation nanoactuators and smart devices.

References

Arefi, Mohammad, Mahmoud Pourjamshidian, Ali Ghorbanpour Arani, and Timon Rabczuk. 2019. "Influence of Flexoelectric, Small-Scale, Surface and Residual Stress on the Nonlinear Vibration of Sigmoid, Exponential and Power-Law FG Timoshenko Nano-Beams." *Journal of Low Frequency Noise, Vibration and Active Control* 38(1):122–42. doi: 10.1177/1461348418815410.

Attia, Mohamed A. 2017. "On the Mechanics of Functionally Graded Nanobeams with the Account of Surface Elasticity." *International Journal of Engineering Science* 115:73–101. doi: 10.1016/j.ijengsci.2017.03.011.

Attia, Mohamed A., and Rabab A. Shanab. 2022. "On the Dynamic Response of Bi-Directional Functionally Graded Nanobeams under Moving Harmonic Load Accounting for Surface Effect." *Acta Mechanica* 233(8):3291–3317. doi: 10.1007/s00707-022-03243-1.

Birman, Victor, and Larry W. Byrd. 2007. "Modeling and Analysis of Functionally Graded Materials and Structures." *Applied Mechanics Reviews* 60(1–6):195–216. doi: 10.1115/1.2777164.

Dangi, Chinika, Roshan Lal, and N. Sukavanam. 2021. "Effect of Surface Stresses on the Dynamic Behaviour of Bi-Directional Functionally Graded Nonlocal Strain Gradient Nanobeams via Generalized Differential Quadrature Rule." *European Journal of Mechanics, A/Solids* 90(March):104376. doi: 10.1016/j.euromechsol.2021.104376.

Dangi, Chinika, and Susmita Naskar. 2025. "Surface Elasticity Based Nonlocal Vibration Analysis of Bidirectional Functionally Graded Tapered Nanobeam." *Advanced Theory and Simulations* 01390. doi: 10.1002/adts.202401390.

Derayatifar, Mahdi, Masuod Tahani, and Hamid Moeenfard. 2017. "Nonlinear Analysis of Functionally Graded Piezoelectric Energy Harvesters." *Composite Structures* 182(April):199–208. doi: 10.1016/j.compstruct.2017.09.030.

El-Galy, Islam M., Bassiouny I. Saleh, and Mahmoud H. Ahmed. 2019. "Functionally Graded Materials Classifications and Development Trends from Industrial Point of View." *SN Applied Sciences* 1(11):1–23. doi: 10.1007/s42452-019-1413-4.

Eringen, A. Cemal. 1983. "On Differential Equations of Nonlocal Elasticity and Solutions of Screw Dislocation and Surface Waves." *Journal of Applied Physics* 54(9):4703–10. doi: 10.1063/1.332803.

Karami, Behrouz, Maziar Janghorban, and Timon Rabczuk. 2020. "Dynamics of Two-Dimensional Functionally Graded Tapered Timoshenko Nanobeam in Thermal Environment Using Nonlocal Strain Gradient Theory." *Composites Part B: Engineering* 182(October 2019):107622. doi: 10.1016/j.compositesb.2019.107622.

Lal, Roshan, and Chinika Dangi. 2021. "Dynamic analysis of bi-directional functionally graded Timoshenko nanobeam on the basis of Eringen's nonlocal theory incorporating the surface effect." *Applied Mathematics and Computation* 395: 125857. https://doi.org/10.1016/j.amc.2020.125857

Li, Hao-Nan, Li Cheng, Shen Ji-Ping, and Yao Lin-Quan. 2021. "Vibration Analysis of Rotating Functionally Graded Piezoelectric Nanobeams Based on the Nonlocal Elasticity Theory." *Journal of Vibration Engineering and Technologies* 9(6):1155–73. doi: 10.1007/s42417-021-00288-9.

Malakooti, Mohammad H., Zhi Zhou, and Henry A. Sodano. 2018. "Enhanced Energy Harvesting through Nanowire Based Functionally Graded Interfaces." *Nano Energy* 52(May):171–82. doi: 10.1016/j.nanoen.2018.07.051.

Marzbanrad, Javad, Salman Ebrahimi-Nejad, Gholamreza Shaghaghi, and Mahya Boreiry. 2018. "Nonlinear Vibration Analysis of Piezoelectric Functionally Graded Nanobeam Exposed to Combined Hygro-Magneto-Electro-Thermomechanical Loading." *Materials Research Express* 5(7). doi: 10.1088/2053-1591/aad0ce.

Parihar, Rityuj Singh, Srinivasu Gangi Setti, and Raj Kumar Sahu. 2018. "Recent Advances in the Manufacturing Processes of Functionally Graded Materials: A Review." *IEEE Journal of Selected Topics in Quantum Electronics* 25(2):309–36. doi: 10.1515/secm-2015-0395.

Penna, Rosa, and Giuseppe Lovisi. 2025. "Surface and Nonlocal Effects in Piezoelectric Nanobeams." *European Journal of Mechanics, A/Solids* 113:105715. doi: 10.1016/j.euromechsol.2025.105715.

Rahmani, Arash, Shirko Faroughi, and Michael I. Friswell. 2020. "The Vibration of Two-Dimensional Imperfect Functionally Graded (2D-FG) Porous Rotating Nanobeams Based on General Nonlocal Theory." *Mechanical Systems and Signal Processing* 144:106854. doi: 10.1016/j.ymssp.2020.106854.

Sahmani, Saeid, and Babak Safaei. 2019. "Nonlinear Free Vibrations of Bi-Directional Functionally Graded Micro/Nano-Beams Including Nonlocal Stress and Microstructural Strain Gradient Size Effects." *Thin-Walled Structures* 140(November 2018):342–56. doi: 10.1016/j.tws.2019.03.045.

Siang, Jee, M. H. Lim, and M. Salman Leong. 2018. "Review of Vibration-Based Energy Harvesting Technology: Mechanism and Architectural Approach." *International Journal of Energy Research* 42(5):1866–93. doi: 10.1002/er.3986.

Singh, Kamalpreet, Kishor Balasaheb Shingare, Tanmoy Mukhopadhyay, and Susmita Naskar. 2023. "Multilevel Fully Integrated Electromechanical Property Modulation of Functionally Graded Graphene-Reinforced Piezoelectric Actuators: Coupled Effect of Poling Orientation." *Advanced Theory and Simulations* 6(4):1–20. doi: 10.1002/adts.202200756.

Wang, Peng, Bai Qiang Liu, Xiao Tong Peng, and Feng Gao. 2025. "Bending and Vibration Behaviour of Functionally Graded Piezoelectric Nanobeams Considering Dynamic Flexoelectric and Surface Effects." *Scientific Reports* 15(1):1–23. doi: 10.1038/s41598-025-98080-2.

Wang, Yuan, Min Hong, Jeffrey Venezuela, Ting Liu, and Matthew Dargusch. 2023. "Expedient Secondary Functions of Flexible Piezoelectrics for Biomedical Energy Harvesting." *Bioactive Materials* 22(August 2022):291–311. doi: 10.1016/j.bioactmat.2022.10.003.

Yang, Tianzhi, Ye Tang, Qian Li, and Xiao-dong Yang. 2018. "Nonlinear Bending, Buckling and Vibration of Bi-Directional Functionally Graded Nanobeams." *Composite Structures* 204(July):313–19. doi: 10.1016/j.compstruct.2018.07.045.

Yu, Pengfei, Weifeng Leng, Liming Peng, Yaohong Suo, and Jinquan Guo. 2021. "The Bending and Vibration Responses of Functionally Graded Piezoelectric Nanobeams with Dynamic Flexoelectric Effect." *Results in Physics* 28:104624. doi: 10.1016/j.rinp.2021.104624.

8

Conclusions and Future Scope

8.1 Overview

The rapid evolution of materials science and computational mechanics continues to open new dimensions for extending nonlocal continuum theories toward the design and analysis of next-generation nanostructures. Despite their promising potential in accelerating the modelling and design of materials, continuum approaches often face functional challenges that can hinder their implementation. In this context, this chapter explores solutions to help mitigate these challenges.

8.1.1 Generalized Mathematical Models

Future investigations should focus on developing generalized models that are based on more advanced mathematical theories, such as fractional differential equations, integral calculus, etc. These frameworks can capture multi-scale interactions more accurately than classical Eringen's theories. Such hybrid models will provide a deeper understanding of nanoscale mechanics, allowing prediction of more accurate nonlocal behaviour.

8.1.2 Stochastic Models and Uncertainty Quantification

Since real-world nanostructures often exhibit imperfections and compositional randomness, introducing stochastic variation in nonlocal elasticity enables probabilistic predictions of mechanical responses. The inclusion of uncertainty quantification techniques, such as polynomial chaos expansion or Bayesian inference, Monte-Carlo, will make the predictive models more robust and reliable for engineering applications.

8.1.3 Multi-Field Modelling of Smart Nanostructures

Future work may explore magneto-electro-thermo-mechanical coupling in the BD-FG nanostructures, enabling the development of self-sensing and adaptive nano systems. Embedding piezoelectric, magnetostrictive,

DOI: 10.1201/9781003583806-8

or ferroelectric layers into BD-FG nanobeams can result in next-generation energy harvesters, nano switches, and reconfigurable resonators capable of in-situ health monitoring and actuation.

8.1.4 AI-Driven Nonlocal Mechanics and Materials Informatics

Inspired by the Materials Genome Initiative (MGI), future studies can integrate machine learning (ML) and deep learning (DL) with nonlocal continuum formulations. Data-driven approaches can accelerate the discovery of structure, optimize gradation profiles, and identify optimal material combinations. Combining nonlocal elasticity solvers with AI-assisted surrogate modelling will enable rapid multiscale predictions without exhaustive computations.

8.2 Summary

This chapter outlines key challenges in mathematical modelling and simulating the nonlocal behaviour of advanced composite nanostructures. One major hurdle—the utilization of more generalized and accurate theory to capture the small-scale effect can be addressed through integral and fractional calculus, which are discussed here. Moreover, the inherent complexity of such mathematical models often requires advanced computational strategies. This work discusses how emerging numerical methods, coupled with publicly available software tools and repositories, can significantly aid in solving and analysing these complex formulations. An important aspect covered is the integration of insights from materials science into these theoretical and computational models. This book holds the potential to revolutionize the design and predictive capabilities of next-generation smart composites.

Index

actuators 1
artificial intelligence (AI) integration 1–2
atomic force microscope (AFM) 1

bi-directionally functionally graded (BD-FG) materials 5, 17, 35
buckling analysis 35, 89

carbon nanotubes (CNTs) 1
Chebyshev collocation method 5
classical continuum theory 2
composite nanostructures 1
computational modelling 1, 6

differential quadrature method (DQM) 5
differential transform method (DTM) 5

elastic foundation models 50
energy relations 21, 37, 51
Eringen's nonlocal elasticity theory (ENT) 20

finite difference method (FDM) 5
finite element method (FEM) 6
functionally graded materials (FGMs) 3
functionally graded nanostructures 1

Galerkin method 5
generalized differential quadrature method (GDQM) 6
Gurtin–Murdoch surface elasticity theory 3

Hamilton's principle 6

in-plane loads 35, 50
Internet of Things (IoT) 1

mathematical modelling 6
micro-electro-mechanical systems (MEMS) 1
modified couple stress theory (MCST) 3
Mori–Tanaka scheme 4

nanobeams 18
nonuniform geometry 78
nanocomposites 1
nanodevices 1
nanostructures 1
nonlocal continuum mechanics 2
nonlocal governing equations 22, 37
nonlocal strain gradient theory (NSGT) 3

Pasternak foundation 50
piezoelectric layers 89
power-law distribution 4

Rayleigh–Ritz method, 5

sensors 1
smart nanostructures 101
Spline interpolation technique 5
strain gradient theory (SGT) 3
surface elasticity theory (SET) 3
surface effects 62

thermal buckling 35
thermo-mechanical vibration 17
Timoshenko beam theory 67

vibration analysis 17, 50, 62
vibration response 50

Winkler foundation 50-59